Contents

Foreword i

List of Acronyms vii

CHAPTER 1: Introduction 1

CHAPTER 2: Getting Started 5
   Preliminary Damage Assessment 5
      What do I need to do to participate in the Preliminary Damage
      Assessment process? 5
   Cost Share 6
   Basic Eligibility 6
   Public Entities 7
   Tribal Governments 7
   Private Nonprofit (PNP) Organizations 7
   The Public Assistance Program Process 7
      What do I need to do to get started? 9
   Getting a Monetary Advance for Your Emergency Work 10
      What do I need to do to receive Immediate Needs Funding? 10

CHAPTER 3: Funding Information You Need to Know 13
   Debris Removal (Category A) 13
   Emergency Protective Measures (Category B) 14
   Permanent Work (Categories C-G) 15
   Management Costs (Declarations on or after November 13, 2007) 17
   Administrative Costs (Declarations before November 13, 2007) 18
      What do I need to do? 19

CHAPTER 4: Kickoff Meeting 21
   Purpose of the Kickoff Meeting 21
      What do I need to do? 21

CHAPTER 5: Funding Options 23
   Infrastructure Restoration Options 23
   Repair or Replacement Projects 23
   Improved Projects 24
   Alternate Projects 26
      What do I need to do? 28

CHAPTER 6: Projects – How They Are Paid 29
Overview..29
Small Projects...29
What do I need to do – small projects? 31
Large Projects...32
What do I need to do – large projects?..33

CHAPTER 7: Subgrant Applications *(Project Worksheets)* and
Cost Estimating..35
General...35
Subgrant Application (*Project Worksheet*) Preparation...........................37
What do I need to do to help develop the Subgrant Applications
(*Project Worksheets*)?...47
Forms..48
Versions/Amendments/Change Orders..48

CHAPTER 8: Dealing With Changes to the Project and
Appealing Decisions 49
Time..49
Cost and Scope...49
Undiscovered and Newly Discovered Damage.......................................50
Appeals 50
What do I need to do? 51

CHAPTER 9: Documentation 53
What do I need to do? 55
Forms..56

CHAPTER 10: Progress Reports, Closeout, and Audit 57
Progress Reports 57
Closeout..57
Audits...58
What do I need to do? 58

CHAPTER 11: Conclusion..61
Sources of Information...61
Apply for Assistance 61
Keep Good Records..61
Your Checklist of Milestone Events 61

APPENDIX A: Private Nonprofit (PNP) Organizations A-1
    Qualifying PNP Facilities A-1
    Qualifying Work A-3
    Small Business Administration A-4
    Some Things You Need to Know A-4
        What do I need to do? A-6

APPENDIX B: Insurance B-1
    Insurance Reductions B-1
    Required Documentation B-2
    Obtaining and Maintaining Insurance B-3
        What do I need to do? B-4

APPENDIX C: Hazard Mitigation C-1
    What It Is C-1
    Some Things You Need to Know C-2
        What do I need to do to obtain Section 406 hazard
        mitigation funding? C-3

APPENDIX D: Historic Preservation D-1
        What do I need to do? D-2

APPENDIX E: Environmental Compliance E-1
        What is my role? E-2
    Form E-3

APPENDIX F: FEMA Forms F-1

APPENDIX G: Glossary of Terms G-1

APPENDIX H: FEMA Policies and Publications H-1

Index I-1

List of Acronyms

ADA	Americans with Disabilities Act
CBRA	Coastal Barrier Resources Act
CBRS	Coastal Barrier Resources System
CEF	Cost Estimating Format
CMF	Case Management File
CFR	Code of Federal Regulations
CWA	Clean Water Act
DHS	Department of Homeland Security
EMAC	Emergency Management Assistance Compact
EO	Executive Order
ESA	Endangered Species Act
FCO	Federal Coordinating Officer
FEMA	Federal Emergency Management Agency
FHWA	Federal Highway Administration
FIA	Federal Insurance Administration
FICA	Federal Insurance Contributions Act
FIRM	Flood Insurance Rate Map
FSR	Final Status Report
GAR	Governor's Authorized Representative
HUD	Department of Housing and Urban Development
ICS	Incident Command System
JFO	Joint Field Office
NEPA	National Environmental Policy Act
NFIP	National Flood Insurance Program
NHPA	National Historic Preservation Act
NMFS	National Marine Fisheries Service
NOAA	National Oceanic and Atmospheric Administration
NRCS	Natural Resources Conservation Service
OCC	FEMA's Office of Chief Counsel
PA	Public Assistance
PAC	PAC Crew Leader

PDA	Preliminary Damage Assessment
PFO	Principal Federal Official
PL	Public Law
PNP	Private Nonprofit
PW	Project Worksheet
RA	Regional Administrator (FEMA)
RPA	Request for Public Assistance
SBA	Small Business Administration
SCO	State Coordinating Officer
SFHA	Special Flood Hazard Area
SHPO	State Historic Preservation Officer
SMD	State Management of Disasters
SOP	Standard Operating Procedure
THPO	Tribal Historic Preservation Officer
USACE	U.S. Army Corps of Engineers
USFWS	U.S. Fish and Wildlife Service

CHAPTER 1

Introduction

When disasters and emergencies occur, the magnitude of work can seem overwhelming. Often, the work is different from the work you usually accomplish, and there is a tremendous amount of it. You must address these events whether Federal assistance is available or not. While FEMA is not able to help you with all of your costs in a Presidentially declared major disaster or emergency, FEMA is able to help with some of them.

The most immediate source to help with response and recovery is your own force account labor, materials, and equipment. They are within your authority and available to you. In a Presidentially declared event, some of your labor, materials, and equipment costs will be eligible for cost-shared FEMA assistance.

Your State may provide labor, materials, equipment, and funds for your response and recovery efforts under State Emergency Plans whether Federal assistance is available or not.

Other jurisdictions and agencies may also come to your aid under mutual aid agreements whether Federal assistance is available or not. If the purpose and provisions of the mutual aid agreements comply with FEMA policy, reasonable costs generally will be eligible for cost-shared FEMA assistance in a Presidentially declared event.

If your needs exceed your local capabilities, you may use contracts to get the work done. Consider Federal, State, and local procurement requirements when procuring goods and services through contracts. If the work is reasonable and necessary and if contracts are awarded according to FEMA requirements, the costs generally are eligible for cost-shared FEMA assistance.

Donations of labor, materials, and equipment can also help with emergency work. If you keep records (hours worked, the work site, description of work, etc.) of what was donated, you may also use such donations toward your portion of any cost-share for other emergency work.

Above and beyond the sources described above, there are other sources you should consider for help in your recovery. They include Federal departments and agencies with specific authority for your damaged facilities (e.g., the U.S. Army Corps of Engineers [USACE], the U.S. Department of the Interior, the Natural Resources Conservation Service [NRCS], the U.S. Department of Transportation Federal Highway Administration [FHWA], and the Departments of Commerce and Housing and Urban Development [HUD]). Your State will help guide you to the appropriate sources for assistance.

When the impact of a major disaster or emergency is so severe that neither the State nor local government can adequately respond, either by direct performance or by contract, the State may request that FEMA supplement State and local efforts with certain emergency work performed directly by a Federal Agency. FEMA, through Mission Assignments, may direct other Federal agencies to perform emergency work or contract for the work as long as the work is eligible under the Stafford Act and Federal regulations. FEMA, other Federal agencies, and contracted entities may provide you with emergency food, water, shelter, medical assistance, search and rescue, debris removal, and other important emergency tasks.

The Federal government shares the cost with you (usually 75 percent Federal and 25 percent non-Federal) according to the FEMA-State Agreement (a formal legal document stating the understandings, commitments, and binding conditions for assistance in the major disaster or emergency) for the major disaster or emergency. When conditions warrant and if authorized by the President, 100 percent Federal funding may even be available for a limited period of time.

FEMA will also share the costs, generally on a reimbursable basis, of the eligible emergency and permanent work, not performed through Mission Assignments, that is part of your response to and recovery from a Presidentially declared event. A summary of the process for obtaining this FEMA assistance is as follows:

Disaster or emergency occurs

Governor requests Federal assistance

Federal and State team collects information on the extent of the damage (Preliminary Damage Assessment)

President declares a major disaster or emergency if the information supports the need for Federal assistance

State briefs potential applicants (Applicants' Briefing) and continues to work with you if you become an applicant

You formally request assistance (Pre-application [*Request for Public Assistance*])

FEMA meets with you and your State Public Assistance (PA) Representative (Applicant Liaison) (Kickoff Meeting)

FEMA staff work with you in defining projects and estimating costs (Project Formulation/Preparation of Subgrant Applications [*Project Worksheets*])

State and FEMA review the projects

FEMA obligates funds for the projects to the State

You work with the State on obtaining the funds[1]

You complete work on your projects

FEMA and State work with you to close out projects and finalize funding (Closeout)

---

1. Applicants should not delay taking the necessary response and recovery actions. Those actions should not be dependent upon receiving Federal funds.

CHAPTERS 2–10 of this Handbook describe this process in more detail. **CHAPTER 11** provides some additional sources of information and gives you a checklist of milestones for obtaining and using FEMA grant funding. The **APPENDICES** are important for understanding requirements for Private Nonprofit (PNP) organizations and special considerations issues (insurance, hazard mitigation, historic preservation, and environmental compliance, including floodplain management). The Appendices also include samples of FEMA forms you may need to complete the Public Assistance process, a Glossary of Terms, and a list of relevant FEMA policies and publications.

Certain individuals will be of help to you in obtaining assistance with your disaster recovery. You will be working with all of them and they are referred to frequently in this Handbook. They include:

State PA Representative Your point of contact, designated by your State, who will help you obtain FEMA assistance.

PA Group Supervisor FEMA's manager of the PA Program. The FEMA Public Assistance Group Supervisor (PA Group Supervisor) will manage the overall operation of PA Program staff, coordinate between PA Program staff, other Federal agencies, and State counterparts, and ensure that the PA Program is operating in compliance with all laws, regulations, and policies.

PAC Crew Leader FEMA's primary point of contact assigned to work with you and guide you through the steps to obtaining funding. The FEMA PAC Crew Leader will advise you on eligibility issues, obtain specialists to assist with your projects, and approve certain project costs.

Project Specialist FEMA's specialist who will work directly with you in assessing your damage sites and in developing scopes of work and cost estimates. The FEMA Project Specialist will also identify the need for other specialists and work with the FEMA PAC Crew Leader in obtaining their services for your projects.

Technical Specialist FEMA's specialists in debris, insurance, cost estimating, hazard mitigation, historic preservation, environmental compliance, floodplain management, and other specialty areas who will help you with special issues.

CHAPTER 2
Getting Started

Preliminary Damage Assessment

After a disaster has occurred and a state of emergency has been declared by your Governor, your State will evaluate the response and recovery capabilities of the State and local governments. If your State believes that response and recovery is beyond the combined capability of the State and local governments, the Governor may request that the FEMA Regional Administrator assist with a Preliminary Damage Assessment.

During this effort, FEMA and State representatives will visit local governments, Tribes, and other potential applicants identified by the State PA Representative to view their damage first-hand and to assess the scope of damage, estimate repair costs, identify unmet needs, and gather information for other management purposes. The Governor will use the information gathered during this Preliminary Damage Assessment process to request Federal assistance to supplement State and local efforts. Occasionally, when the catastrophic nature and magnitude of an event are very clear, the President may declare a major disaster or emergency immediately, thereby abbreviating the Preliminary Damage Assessment.

What do I need to do to participate in the Preliminary Damage Assessment process?

- Contact your State PA Representative to report that your jurisdiction has suffered damages.

- Because the situations are generally urgent, the FEMA/State team has limited time to collect the damage information. To help document the magnitude and impact of the disaster or emergency and to help the Governor obtain Federal assistance, be prepared to show the team a representative sampling of your damage sites.

- Provide a map (preferably 8½ x11 inches) of your jurisdiction showing all damage sites.

- Inform the Preliminary Damage Assessment team of your emergency work (along with any associated immediate expenditures) and by helping to estimate the costs to repair damaged facilities.

- You also can assist the Preliminary Damage Assessment team by estimating the costs you have incurred. See **CHAPTER 7** for more information on cost estimating.

Cost Share

After a Presidential declaration has been made, FEMA will publish in the Federal Register the President's decision on which areas are eligible for assistance and the types of assistance available. The Federal share will always be at least 75 percent of the eligible costs and may be more, depending on the severity of the disaster or emergency. All applicants, including PNP organizations, are subject to the cost share outlined in the FEMA-State Agreement. This cost sharing applies to all eligible work. The FEMA-State Agreement will also describe the cost share provisions that apply to Direct Federal Assistance (for example, Mission Assignments) provided by a Federal agency, including any administrative costs of the Federal agency's assistance.

Basic Eligibility

Once a Presidential declaration has been made, damage that meets all of the following criteria may be eligible for cost-shared FEMA assistance:

- The damage is a direct result of the declared event;
- The damage that occurs can be tied directly to the declared event;
- The damage occurred within the designated disaster area (except sheltering and evacuation activities that may be located outside the designated area);
- The damage is the applicant's legal responsibility at the time of the disaster;
- With very few exceptions, the damage occurred at a facility in active use; and
- The damage is not within the specific authority of another Federal program.

Assistance can be for debris removal, emergency protective measures, and permanent restoration of your damaged infrastructure. There are other eligibility requirements, but the first test of eligibility is the above list.

Public Entities

State and local government entities are eligible applicants. This group includes: the States of the United States of America, the District of Columbia, the territories of Guam, the Virgin Islands, and American Samoa, the Commonwealths of Puerto Rico and the Northern Mariana Islands, counties, municipalities, cities, towns, townships, local public authorities, school districts, special districts, intrastate districts, councils of government, regional or interstate government entities, and agencies or instrumentalities of a local government.

Tribal Governments

Federally recognized Indian Tribal governments, including Alaska Native villages and organizations, are eligible applicants. Privately owned Alaska Native Corporations are not. Generally, Indian Tribes are considered applicants and receive grant funds through the State.

In some States, however, State regulations prohibit the State from acting as grantee for FEMA funds for an Indian Tribe. In such cases, or upon the choice of the Tribal government, the Tribal government may act as its own grantee. The Tribal government must apply to the FEMA Regional Administrator to become its own grantee. An Indian Tribal government that chooses to act as its own grantee becomes responsible for the entire non-Federal share of the public assistance grant. If the Indian Tribal government is the grantee, it takes on the roles and responsibilities described throughout this Handbook for the State.

Private Nonprofit (PNP) Organizations

PNP organizations that own or operate facilities that provide certain services of a governmental nature are eligible applicants. (Further discussion of the eligibility of PNPs is provided in **APPENDIX A**.)

The Public Assistance Program Process

Overview of the Process. The Applicants' Briefing begins your formal participation in the Public Assistance Program. What follows will be submittal of your Pre-application (*Request for Public Assistance*) (**CHAPTER 2**); a Kickoff Meeting (**CHAPTER 4**), a meeting in which your FEMA PAC Crew Leader will guide you through the process of getting help;

development of Subgrant Applications (*Project Worksheets*) (**CHAPTER 7**); and, eventually, closeout (**CHAPTER 9**). Generally, you will have 6 months to complete your debris clearance and emergency work, and 18 months to complete your permanent restoration work. See **CHAPTER 6** for information on how projects are funded, **CHAPTER 8** for more timeline information, and the checklist in **CHAPTER 11** for important milestones associated with the process.

Applicants' Briefing. One of the first events in the process of getting assistance with damages is the Applicants' Briefing. It is a meeting conducted by the State to provide an overview of the Public Assistance Program. The State will notify potential applicants (State and local governments, Tribal governments, and PNP organizations) of the date, time, and location of this meeting as soon as practicable after the President's declaration. To obtain maximum benefit from the information presented at the briefing, you should send representatives of your organization's management, emergency response, public works, and accounting/finance/procurement operations.

During the briefing, a State PA Representative will tell you about:

- The disaster or emergency that was declared
- The incident period for which damages may be eligible
- Applicant, facility, work, and cost eligibility
- Eligibility of PNP organizations and facilities (see also **APPENDIX A**)
- The process for documenting projects
- Funding options
- Federal procurement standards
- Record keeping and documentation requirements
- Special Considerations
    - Insurance claims and requirements (see also **APPENDIX B**)
    - Hazard Mitigation (see also **APPENDIX C**)
    - Historic preservation compliance requirements that apply to the receipt of Federal assistance (see also **APPENDIX D**)

- Environmental compliance requirements, including floodplain management, that apply to the receipt of Federal assistance (see also **APPENDIX E**)

■ State grant requirements

FEMA personnel will attend the meeting to clarify any issues regarding the Public Assistance Program. You should submit your Pre-application (*Request for Public Assistance*) (FEMA Form 90-49) at this meeting. This is FEMA's official application form to apply for the Public Assistance Program. It is a simple, short form with self-contained instructions. The Pre-application form (*Request for Public Assistance*) asks for general contact information, which FEMA and the State use to start the grant process and open your Case Management File. Your Case Management File will contain general claim information as well as records of meetings, conversations, phone messages, and any special issues or concerns that may affect funding for your projects.

What do I need to do to get started?

- You need to submit your Pre-application form (*Request for Public Assistance*) to your State PA Representative within 30 days of the date of designation of your disaster area. If you haven't already prepared it prior to the Applicants' Briefing, complete it and submit it online (if your State provides for that) or during the meeting. Don't wait until you have identified all your damage before submitting your Pre-application form (*Request for Public Assistance*), as this document starts the process for getting your funding.

- In addition to submitting your Pre-application form (*Request for Public Assistance*), be sure to raise any concerns or identify any needs you may have (e.g., for getting a monetary advance for emergency work) to the State PA Representative conducting the briefing.

- If you are a recognized Indian Tribal government, confer with the relevant State representatives and FEMA to determine if you will be managing your own grant or applying as an applicant under a State grant.

- You may obtain a copy of a Pre-application form (FEMA Form 90-49) (*Request for Public Assistance*) from the State emergency management organization, at the Applicants' Briefing, or through FEMA's Web site at **www.fema.gov/government/grant/pa/forms.shtm**.

Getting a Monetary Advance for Your Emergency Work

Immediate Needs Funding. When conducting the Preliminary Damage Assessment, FEMA and the State identify any circumstances in which there are immediate needs. If a major disaster or emergency is declared by the President, and the State determines that damage costs require immediate cash flow, the State may request Immediate Needs Funding. Up to 50 percent of the Federal share estimate for emergency work to be completed within 60 days will then be placed in the State's account. The State is responsible for disbursement. Because this money can be made available in advance of the normal Public Assistance Program process following the declaration of a major disaster or emergency, paperwork and processing times are reduced and you may receive emergency funds sooner. Your jurisdiction must have participated in the Preliminary Damage Assessment and be included in the President's declaration to qualify for Immediate Needs Funding.

This funding is for the most urgent work during the response to, or immediately after a major disaster or emergency. The funds may be provided to any eligible applicant for eligible emergency work performed immediately and paid for within the first 60 days following declaration. The funding is for emergency work (i.e., debris removal and emergency protective measures). Additional discussion of emergency work is provided in **CHAPTER 3**. Immediate Needs Funding can also be used for overtime payroll, equipment costs, and materials purchases, and contracts when these costs are incurred for emergency work. Immediate Needs Funding is not intended for permanent work, work that involves environmental or historic concerns, work covered by insurance, large debris or demolition projects that require more than 60 days to complete, or debris removal and disposal within the Coastal Barrier Resources System. Any Immediate Needs Funding you receive will be offset by the costs of your actual emergency work projects as the Subgrant Applications (*Project Worksheets*) are received. Your eligible permanent work costs will not be obligated until your Immediate Needs Funding is reimbursed. The funding for eligible permanent work costs is requested through the Subgrant Application (*Project Worksheet*) process.

What do I need to do to receive Immediate Needs Funding?

- ▶ If your damages have been surveyed in the Preliminary Damage Assessment, you may be eligible for Immediate Needs Funding.

- Your State will notify you about how to apply for Immediate Needs Funding. Typically the State will ask you to send a letter of request to a designated State official. In addition, you must submit a completed Pre-application (*Request for Public Assistance*) (FEMA Form 90-49) before the State will release any Immediate Needs Funding.

- You will need documentation to support the costs you claim under Immediate Needs Funding procedures. Failure to document costs will result in loss of the funding.

Expedited Payment. The Expedited Payment is similar to Immediate Needs Funding. Though the State requests Immediate Needs Funding on behalf of all applicants, an individual applicant may request an Expedited Payment on its own. The State will need to process your request and disburse the funds that FEMA obligates.

Expedited Payments are made for applicants who participated in the Preliminary Damage Assessment and who file a Pre-application (*Request for Public Assistance*). FEMA will obligate 50 percent of the Federal share of the estimated costs of work under Categories A and B (Debris Removal and Emergency Protective Measures) as estimated during the Preliminary Damage Assessment. FEMA will make payment for Debris Removal within 60 days after the estimate was made and no later than 90 days after the Pre-application (*Request for Public Assistance*) was submitted. Because the money can be made available on the basis of an estimate, paperwork and processing times are reduced and you can receive emergency funds sooner than if you followed the normal process. Expedited Payments are possible only if your county has been included in the President's declaration.

Expedited payments are not intended for work that involves environmental or historic concerns or work that is covered by insurance. Any Expedited Payment you receive will be offset by the costs of your actual emergency work Subgrant Applications (*Project Worksheets*) as they are received. As with Immediate Needs Funding, your eligible permanent work costs will not be obligated until your Expedited Payment is reimbursed. The funding for eligible permanent work costs is requested through a Subgrant Application (*Project Worksheet*).

CHAPTER 3

Funding Information You Need to Know

FEMA can help you with your costs for debris removal, emergency protective measures, and infrastructure restoration. A summary of eligibility follows, but you will need to work with the FEMA PAC Crew Leader and the State PA Representative assigned to you to determine if your facilities, work, and costs meet Public Assistance Program eligibility criteria. You can obtain more information on eligibility from your FEMA PAC Crew Leader and on the FEMA Web site **www.fema.gov/government/grant/pa/**. Your FEMA PAC Crew Leader can also advise you on the special eligibility requirements of PNP facilities (see also **APPENDIX A**).

Debris Removal (Category A)

FEMA can assist you with funding for the clearance, removal, and/or disposal of items such as trees, woody debris, sand, mud, silt, gravel, damaged building components and contents, wreckage produced during the conduct of emergency work, and other disaster-related wreckage. For debris removal to be eligible, the work must be necessary to:

- Eliminate an immediate threat to lives or public health and safety

- Eliminate immediate threats of significant damage to improved public or private property when the measures are cost effective

- Ensure the economic recovery of the affected community to the benefit of the community-at-large

- Mitigate the risk to life and property by removing substantially damaged structures as needed to convert property acquired using FEMA hazard mitigation program funds to uses compatible with open space, recreation, and wetland management practices

Debris removal is the responsibility of government agencies. If necessary, your FEMA PAC Crew Leader can advise you on the few situations in which debris removal by PNP organizations may be eligible.

Removal of debris on public property that is required to allow continued safe operation of governmental functions or to alleviate an immediate threat is generally eligible. Debris on private property rarely meets the public interest standard because it does not affect the public-at-large and most often is not the legal responsibility of a State or local government. Debris removal from private property is usually the responsibility of the individual property owner.

Emergency Protective Measures (Category B)

FEMA can also help pay for actions taken by the community (almost always government agencies) before, during, and after a disaster to save lives, protect public health and safety, and prevent damage to improved public and private property. Examples of measures that may be eligible include:

- Warning of risks and hazards
- Search and rescue
- Emergency evacuations
- Emergency mass care
- Rescue, evacuation, transportation, care, shelter, and essential needs for humans affected by the outbreak and spread of an influenza pandemic
- Protection for an eligible facility
- Security in the disaster area
- Provision of food, water, ice, and other essential items at central distribution points
- Temporary generators for facilities that provide health and safety services
- Rescue, care, shelter, and essential needs for household pets and service animals if claimed by a State or local government
- Temporary facilities for schools and essential community services
- Emergency operations centers to coordinate and direct the response to a disaster
- Demolition and removal of public and private buildings and structures that pose an immediate threat to the safety of the general public

- Removal of health and safety hazards
- Construction of emergency protection measures to protect lives or improved property (for example, temporary levees)
- Emergency measures to prevent further damage to an otherwise eligible facility (for example, boarding windows)
- Restoration of access
- Inspections if necessary to determine whether structures pose an immediate threat to public health or safety

Permanent Work (Categories C-G)

FEMA can also help pay to restore facilities through repair or restoration, to pre-disaster design, function, and capacity in accordance with codes or standards.

- **Roads and Bridges (Category C).** Roads (paved, gravel, and dirt) are eligible for permanent repair or replacement, unless they are Federal-Aid roads (which are supported by the FHWA). Eligible work includes repair to surfaces, bases, shoulders, ditches, culverts, low water crossings, and other features, such as guardrails. Repairs necessary as the result of normal deterioration, such as alligator cracking or rotted timbers, are considered pre-disaster damage and not eligible. Eligible work for bridges includes decking and pavement, piers, girders, abutments, slope protection, and approaches.

- **Water Control Facilities (Category D).** Permanent repairs are not eligible for flood control works and federally funded shore protective devices since the primary authority for the restoration of those facilities is with the U.S. Army Corps of Engineers and the National Resources Conservation Service. But FEMA can help with permanent repairs for other water control facilities, such as those that were built for channel alignment, recreation, navigation, land reclamation, maintenance of fish and wildlife habitat, interior drainage, irrigation, and erosion prevention. (See **APPENDIX A** for special requirements for PNP irrigation facilities.)

- **Buildings and Equipment (Category E).** Buildings, structural components, interior systems (e.g., electrical and mechanical systems), building contents, vehicles, and equipment are eligible for repair or replacement. Replacement of pre-disaster quantities of consumable supplies and inventory, the replacement of library books and publications, and the stabilization (but not re-creation from original sources) of damaged files are also eligible. If disaster-related mud, silt, or other accumulated debris does not pose an immediate threat but its removal is necessary to restore the building, its removal is eligible as permanent work; if it does pose an immediate threat, the disaster-related work will fall under Category A (Debris Removal).

- **Utilities (Category F).** The repair or restoration of utilities is also eligible. Utilities include:

    ▶ Water treatment plants and delivery systems

    ▶ Power generation and distribution facilities, including natural gas systems, wind turbines, generators, substations, and power lines

    ▶ Sewage collection systems and treatment plants

    ▶ Communications

    You are responsible for determining the extent of damage to utility systems. General surveys to look for damage are not eligible, but if you discover damage, the inspection of the damaged section is eligible. Any increased operating expenses resulting from the disaster or lost revenue are not eligible; however, the cost of establishing temporary emergency utility services in the event of a shut-down may be eligible as emergency work.

- **Parks, Recreational Areas, and Other Facilities (Category G).** Publicly owned facilities in this category are generally eligible. They include:

    ▶ Playground and picnic equipment

    ▶ Swimming pools, golf courses, and tennis courts

    ▶ Piers

    ▶ Some beaches (you will need to work with your FEMA Project Specialist to determine if yours meet the criteria for assistance)

    ▶ Mass transit facilities, such as rail systems

    ▶ Facilities that do not fit in Categories C-F, such as fish hatcheries

    ▶ Supporting facilities (e.g., roads, buildings, and utilities) that are located in parks and recreational areas, subject to the eligibility criteria for Categories C, D, E, and F

Natural features, such as the banks of streams, are not facilities and are not eligible for repair. In addition, the replacement of trees, shrubs, and other ground cover is not eligible for any facility in any category of work. This means that replacement of grass and sod (including for recreational and sports areas) generally is not eligible for the Public Assistance Program. The one exception is grass or sod required as part of another measure to stabilize a slope and minimize erosion.

Recreational areas owned by PNP organizations are not eligible for assistance.

- **Fire Management (Category H).** Fire Management Assistance declarations are approved by FEMA's Recovery Directorate Assistant Administrator, or his/her designee, in response to a State's request for fire management assistance. FEMA's Fire Management Assistance Grant Program is authorized by the Stafford Act and is available to States, local governments, and Indian Tribal governments. It is intended to aid with the mitigation, management, and control of fires burning on publicly- or privately-owned forests or grasslands that would threaten such destruction as would constitute a major disaster. The grants cover fire-related activities, such as firefighting and support services, pre-positioning Federal, out-of-state, and international resources for up to 21 days, evacuations, sheltering, traffic control, emergency operations centers, and temporary repairs of damage caused by firefighting activities. More information on the grants can be obtained at **www.fema.gov/government/grant/fmagp/index**.

Management Costs (Declarations On or After November 13, 2007)

The Stafford Act provides that the State may request reimbursement for its management and administration of disaster assistance, as well as yours. The maximum amount of funding that can be requested from FEMA is a set percentage of the Federal share of assistance granted for the major disaster or emergency. For events declared on or after November 13, 2007, the initial management cost rate is set at 3.34 percent for major disaster declarations and 3.9 percent for emergency declarations. These rates cover management costs of both the State and the applicants. The State determines the portion distributed to you. FEMA will review the two rates no later than November 13, 2010.

The allowance covers any indirect cost, any administrative expense, and any other expense for the major disaster or emergency that is not directly chargeable to a specific project. The State will not receive any separate funding for State Management Administrative Costs and will not be reimbursed for State indirect costs. Records need to be kept on how the funds were spent and any surplus is subject to reimbursement to FEMA. In addition to management costs, FEMA will reimburse direct administrative costs incurred by the State and applicants that are properly documented and directly chargeable on a Subgrant Application (*Project Worksheet*) for a specific project.

Administrative Costs (Declarations Before November 13, 2007)

For major disasters or emergencies declared before November 13, 2007, the allowance described in the *Public Assistance Guide* (FEMA 322) and *Public Assistance Policy Digest* (FEMA 321) for the State and applicants will continue to apply.

- The administrative allowance described there is intended to meet your costs for administering your funding. It covers your direct and indirect costs incurred in requesting, obtaining, and administering Public Assistance funding (e.g., identifying damage, attending the Applicants' Briefing, completing the Pre-application [*Request for Public Assistance*], preparing Subgrant Applications [*Project Worksheets*], establishing files, providing documentation, assessing damage, collecting cost data, developing cost estimates, working with the State during project monitoring, final inspections, and audits, and preparing for audits).
- It is not intended to cover direct costs of managing specific projects because these costs generally are eligible as part of the grant for each project.
- The administrative allowance is calculated as a percentage of total eligible costs that are approved for you in a given major disaster or emergency. The percentage is calculated using the same sliding scale used for the State's administrative allowance.
- For expenditures beyond eight years from a major disaster declaration date or beyond two years from an emergency declaration date, special restrictions apply.
- Records need to be kept on how the funds were spent and any surplus is subject to reimbursement to FEMA.

What do I need to do?

- Identify your damages and work with your State PA Representative to request assistance with your costs.

- For major disasters or emergencies declared before November 13, 2007, FEMA automatically calculates your allowance for administrative costs when processing your Subgrant Applications (*Project Worksheets*) and forwards your allowance to the State.

- For major disasters or emergencies declared on or after November 13, 2007, you will need to work with your State PA Representative to establish the amount and method of delivery of your allowance for management costs. Keep documentation on how the funds were spent because funding in excess of costs you can document must be returned to FEMA.

CHAPTER 4

Kickoff Meeting

Purpose of the Kickoff Meeting

The Kickoff Meeting is a substantive, project-oriented meeting for you, your State PA Representative, and FEMA. It differs from the Applicants' Briefing conducted by the State at the onset of disaster operations. While the Applicants' Briefing gives a general overview of the Public Assistance Program and description of the application process, this meeting is conducted specifically for you by your FEMA PAC Crew Leader in order to provide a much more detailed review of the Public Assistance Program and an opportunity to discuss your specific needs. It will include eligibility and documentation information that pertains to your organization. You will also meet the FEMA Project Specialist who will be working with you on your projects at this meeting.

The FEMA PAC Crew Leader will also discuss Special Considerations (i.e., insurance, hazard mitigation opportunities, and compliance with historic preservation and environmental laws, including floodplain management) that could potentially affect the type and amount of assistance available and the documentation needed. For this reason, Technical Specialists will be included in the meeting if Special Considerations issues are anticipated. (See **APPENDICES B, C, D, and E** for further information on Special Considerations.) Satisfactorily addressing Special Considerations on each project is critically important, as failure to do so may jeopardize the funding of your project.

The Kickoff Meeting is the first of what will be many project-oriented contacts among you, your State PA Representative, and FEMA. By the end of the meeting, you will know what to expect and will receive the information you need to proceed with the Public Assistance Program process. This is a good opportunity to request clarification of anything you do not understand and raise any issues that concern you.

What do I need to do?

► You can expect to be contacted by your FEMA PAC Crew Leader or State PA Representative within one week after submission of your Pre-application form (*Request for Public Assistance*). If you have not been contacted after one week, call your State PA Representative to arrange the first meeting.

- In preparation for the meeting, compile your list of damages by location. If you can, include a description of the damage and scope of work, the nature of the work (i.e., debris removal, emergency measures, or permanent restoration), estimated cost, whether you will be completing the work with your own force account resources or by contract, whether the facility was damaged in a prior disaster, and whether you are interested in hazard mitigation measures that may prevent future damage.

- Identify circumstances that require special review, such as insurance coverage, environmental issues, floodplain management issues, and historic preservation concerns. The earlier these conditions are known, the faster they can be addressed; and they must be addressed before project funding can be approved.

- Bring your insurance policies and settlement documents.

- Be prepared to identify your priority projects.

- Be prepared to discuss your need for Expedited Payments or Immediate Needs Funding, if you have received either or have decided you need one or the other. Cost estimates with justification and supporting documentation will be necessary, and you will need to submit documents to account for all incurred costs.

- Consider inviting those from your organization with a working knowledge of the needed repairs and emergency work costs, your public works officials who will be managing the repairs, a procurement official, a financial management official, a risk manager who is familiar with your insurance coverage, and the individual(s) who will be keeping progress and financial records on each of the projects.

CHAPTER 5

Funding Options

Infrastructure Restoration Options

You have several options regarding the restoration of your damaged facility. They include the following possibilities:

- Repair the facility to current codes and standards. (Repair)

- Repair the facility and include FEMA-funded hazard mitigation measures. (Repair)

- Pay for improvements yourself as you repair the facility either on its current site or at a new location. (Improved Project)

- Repair the facility to a code or standard that exceeds FEMA-approved codes and standards. (Improved Project)

- Replace the facility because it is damaged beyond repair. (Replacement)

- Relocate the facility out of a floodplain. (Replacement)

- Build a large facility to replace several damaged facilities in separate locations. (Improved Project)

- Do not repair a damaged facility that is no longer needed and use the funds for other purposes. (Alternate Project)

- Use some of the Public Assistance funds to make the damaged facility safe and secure (repairing it later on your own), and use the balance for other eligible projects. (Alternate Project)

Repair or Replacement Projects

FEMA provides assistance to help you restore an eligible facility to its pre-disaster design, function, and capacity. As part of the basic restoration, FEMA may also pay for upgrades that are necessary to meet the requirements of reasonable applicable codes and standards and may pay for reasonable and cost-effective hazard mitigation measures as part of the repair. PNP organizations have some special provisions for their eligibility. (See **APPENDIX A** for more information about PNP Organizations.)

If a facility is damaged to the point where you think it should be replaced rather than repaired, the following calculation, known as the 50 Percent Rule, is used to determine whether replacement is eligible.

Your FEMA PAC Crew Leader will guide you through your eligible costs and the calculation to determine whether your facility is damaged beyond repair.

> **50 Percent Rule**
>
> If the repair cost divided by the replacement cost is less than 0.5, then only the repair cost is eligible.
>
> If the repair cost divided by the replacement cost is more than or equal to 0.5, then the replacement cost is eligible.

Your FEMA PAC Crew Leader will also be able to guide you through your decision to improve your facility as you restore it or even to not repair your facility at all.

Improved Projects

Most applicants use FEMA funds to simply return their damaged facilities to their pre-disaster function and capacity. But the repair of disaster damage is also an opportunity to improve on the facility that was damaged. If you want to do this, you must use your own funds for the improvements and FEMA funds for the basic repair. Examples of improved projects include:

- Laying asphalt on a gravel road
- Replacing a firehouse that originally had two bays with one that has three
- Increasing the capacity or changing the design of a school

FEMA will provide the funding for the FEMA share of the eligible costs for repairing your facility. Additional costs are your responsibility.

If you are contemplating an improved project, here are some things to keep in mind:

- The improved project must have the same function and at least the pre-disaster capacity as that of the pre-disaster facility to be eligible.

- Time limits that would be associated with repairing the damaged facility to its pre-disaster design apply to the improved project construction. **CHAPTER 8** includes additional discussion of time limits.

- **Funding for improved projects is capped at the Federal share of the costs that would be associated with repairing or replacing the damaged facility to its pre-disaster design, or to the actual costs of completing the improved project, whichever is less.** For the most part, funding for approved work cannot be tracked within the improved project because of physical changes or contracting arrangements. However, if eligible repair or replacement costs exceed the original estimate and costs can be documented (i.e., you were to track eligible costs separately from improvement costs), you may appeal for additional funds. **CHAPTER 8** includes additional discussion of appeals.

- You must obtain approval for an improved project from the State prior to the start of construction. Any improved project that results in a significant change from the pre-disaster configuration (i.e., different location, footprint, function, or size) of the facility must also be approved by FEMA prior to construction to ensure completion of the appropriate historic preservation and environmental compliance reviews.

- Any additional costs for complying with codes and standards or compliance with environmental and historic preservation laws, regulations, and Executive Orders (EOs) required by the construction of the improvements, but not required by the eligible scope of work for repair, are not eligible.

- Funds to construct the improved project can be combined with a grant from another Federal agency and/or a FEMA-approved alternate project; however, Federal grants cannot be used to meet the State or local cost-share requirement unless the legislation for the other grant allows such use (e.g., the Community Development Block Grant program).

- If the original facility is being repaired and improvements are being added, FEMA may provide assistance with hazard mitigation measures under Section 406 of the Stafford Act. These funds must be applied to the original facility. Improved projects that involve a new facility on the same site or on a different site are not eligible for Section 406 Hazard Mitigation funding.

Alternate Projects

Occasionally, an applicant may determine that the public welfare is not best served by restoring the facility or the function using FEMA funds. This usually occurs when the service provided by the facility is no longer needed, even though the facility may have been in use at the time of the disaster. If this happens, you may apply through the State to FEMA to use the eligible funds for your damaged facility for an alternate project, that is, another public purpose. FEMA will estimate all eligible costs of repairing your facility, evaluate your proposal to use the funds for another purpose, and obligate the FEMA share of an approved proposal.

Examples of alternate projects include:

- Repair or expansion of other public facilities
- Construction of new public facilities
- Demolition of the original structure
- Purchase of capital equipment
- Funding of cost-effective hazard mitigation measures in the area affected by the disaster
- Funding project shortfalls due to mandatory National Flood Insurance Program (NFIP) reductions on buildings in floodplains
- Supplemental funds for an improved project

If you are contemplating an alternate project, keep the following in mind:

- Funds may be used on more than one alternate project.
- An alternate project option may be proposed only for permanent restoration projects within the declared disaster area.
- Funds for emergency work (i.e., debris removal and emergency protective measures) cannot be used for alternate projects.

- Alternate projects for governmental entities are eligible for 90 percent of the approved Federal share of the estimated eligible costs associated with repairing the damaged facility to its pre-disaster design, or 90 percent of the approved Federal share of the actual costs of completing the alternate project, whichever is less. For example:

$100,000	*estimate of eligible damage*
x .75	*adjustment to account for percentage of Federal cost share*
$75,000	*the Federal share*
x.90	*10 percent reduction for alternate project funding*
$67,500	*maximum grant amount you can receive*

    You must spend at least $75,000 on approved alternate projects to receive $67,500 in Federal funds.

- Alternate projects for PNP organizations are eligible for 75 percent of the approved Federal share, or 75 percent of the approved Federal share of the actual costs of completing the alternate project, whichever is less. The alternate project must also be one that would be eligible for Public Assistance funding in a subsequent disaster. The calculation would be similar to the one above for public entities with the exception that the $75,000 would be multiplied by .75 (instead of .90) to create the 25 percent reduction for alternate projects of PNP organizations and the maximum amount of Federal funds would be $56,250. See **APPENDIX A** for other eligibility provisions for PNP organizations.

- At a minimum, your original facility must be rendered safe and secure or demolished. If you opt to keep the damaged facility for a later or other use, it will not be eligible for Public Assistance Program funding in a subsequent disaster unless it is repaired.

- The costs of complying with laws, regulations, and Executive Orders on the damaged facility are considered project costs for purposes of calculating the grant. Executive Orders are legally binding orders issued by the President to Federal Administrative Agencies. However, any additional costs for complying with codes and standards or compliance with environmental and historic preservation laws, regulations, and Executive Orders for the alternate facility are not eligible.

- The proposed alternate project may not be located in the regulatory floodway and will have to be insured if located in the 100-year floodplain. Funding may not be used for operating costs or to meet the State or local share requirement on other Public Assistance projects or projects that utilize other Federal grants unless the legislation for the other grant allows such use (e.g., the Community Development Block Grant program).
- Section 406 (Stafford Act) hazard mitigation funds cannot be applied to an alternate project.
- FEMA must ensure that the proposed projects are an appropriate use of funds and comply with environmental and historic preservation laws.

What do I need to do?

- Assist with the completion of the Subgrant Application (*Project Worksheets*) for your damaged facilities, even if you have not yet decided whether you might opt for an improved or alternate project. You and FEMA need to establish your eligible disaster damages so that you will know the amount of funds potentially available to you.
- Decide early whether or not you intend to restore your damaged facility and, if not, how you want to use the eligible funds.
- Discuss your options with your State PA Representative and your FEMA PAC Crew Leader.
- You must obtain approval for an alternate project from the State prior to the start of construction. Any alternate project that results in a significant change from the pre-disaster configuration (that is, different location, footprint, function, or size) of the facility must also be approved by FEMA prior to construction to ensure completion of the appropriate historic preservation and environmental compliance reviews.
- You must submit any proposal for an alternate project through the State to FEMA. FEMA approval is required for **all** alternate project proposals. **Do not start demolition or construction without FEMA approval.**

CHAPTER 6
Projects – How They Are Paid

Overview

Under the Public Assistance Program, FEMA provides eligible grant funds to the State. The State then provides the Federal share of funding to you for your projects from the overall grant.

Your projects are categorized as either large or small, as determined by the estimated cost of eligible damages. This separation facilitates project review, approval, and funding. The cost threshold that distinguishes large from small projects changes annually and is published in the Federal Register. The threshold is $63,200 for the fiscal year ending September 30, 2010.

- Small Projects < $63,200
- Large Projects ≥ $63,200

Small project funding is based on estimated costs, if actual costs are not yet available. Payment is final, made on the basis of the initial approved amount, whether estimated or actual. Subgrant Applications (*Project Worksheets*) are not revised unless there are gross errors, omissions, or changes in scope; this occurs very rarely. If they are revised, they may change in funding and/or scope of work.

Large project funding is based on documented actual costs. Because of the complexity and nature of most large projects, work is typically not complete at the time of FEMA approval. Therefore, most large projects are initially approved based on estimated costs. Funds generally are made available to you on a progress payment basis as you complete work.

Small Projects

Payment for small projects is made at the time of project approval on the basis of the estimate. The State is required to make payment of the Federal share to you as soon as practicable after FEMA has obligated the funds. The advantage of the small project process is that funds are available as the Subgrant Applications (*Project Worksheets*) are approved, rather than reimbursed after you submit documentation of incurred costs.

Because small projects are likely to be similar to work you do on a regular basis, you may want to expedite processing further by preparing the Subgrant Applications (*Project Worksheets*) for your small projects on your own. If you do, FEMA will provide guidance, validate a sample, and then obligate funds based on a successful validation. As an alternative, you may ask your FEMA Project Specialist to complete your Subgrant Applications (*Project Worksheets*).

The funding for individual small projects is fixed, regardless of the final costs incurred. Generally, FEMA does not perform an inspection of completed small projects; however, the State must certify that the applicant completed the work in compliance with all applicable laws, regulations, and policies. Therefore, the State may decide to review some or all of your small projects.

If you spend less than the amount FEMA approved, the Federal share will not be reduced to match actual costs. However, if you incur costs significantly greater than the total amount approved for all of your small projects, you may appeal for additional funding. Note that this opportunity applies only to a significant net cost overrun for all of your small projects, not to an overrun for an individual project. If you request additional funds, it must be done as an appeal and submitted within 60 days of completion of your last small project. You should be prepared for a complete audit of all your small projects if you request additional funds.

Small project validation. If you prepare your own Subgrant Applications (*Project Worksheets*), FEMA will conduct a validation of the projects to confirm that you understand the documentation and eligibility provisions of the Public Assistance Program and that you are capturing all eligible costs. Your records are the basis for validation.

The sample validated will be limited to the minimum amount of review needed to ensure statutory and regulatory compliance. If the projects are submitted within the first 60 days after the Kickoff Meeting, this will be 20 percent of all your small projects. For projects submitted after 60 days, 100 percent will be validated. Validation can normally be completed within 15 days of submission of all Subgrant Applications (*Project Worksheets*) to the FEMA PAC Crew Leader.

The FEMA PAC Crew Leader will review each worksheet to ensure the scope of work is complete and that all Special Considerations have been identified. The FEMA PAC Crew Leader, with input from the State, will select two samples from all the small projects you have submitted. Each sample is made up of 20 percent of the total small projects. If the first sample does not pass validation, FEMA will conduct a second validation using the remaining sample. If you have four or fewer Subgrant Applications (*Project Worksheets*), a minimum of one Subgrant Application (*Project Worksheet*) will always be validated.

Validation, which is generally done by your FEMA Project Specialist, includes the following steps:

- Visit the sites to confirm all aspects of the project description are accurate and complete and that all Special Considerations have been identified.

- Confirm the description of damage and dimensions is complete, accurate, and eligible.

- Confirm the scope of work is complete, accurate, and eligible.

- Review available actual cost records to ensure completeness, accuracy, and eligibility.

- Review your estimated costs to ensure completeness, accuracy, reasonableness, and eligibility.

Small projects with Special Considerations (i.e., insurance, hazard mitigation measures, historic preservation, and environmental compliance, including floodplain management) will be individually funded as the Special Considerations issues are resolved. See **APPENDICES B, C, D, AND E**.

What do I need to do – small projects?

- ► You can expedite the handling of your projects by identifying any Special Considerations (i.e., insurance, hazard mitigation, historic preservation, and environmental compliance, including floodplain management issues) that relate to your project.

- ► If you do not want to prepare your own Subgrant Applications (*Project Worksheets*), notify your FEMA PAC Crew Leader so that FEMA will know to prepare them for you.

- ► If you are preparing your own Subgrant Applications (*Project Worksheets*), notify the FEMA PAC Crew Leader when you have submitted all, or a large batch, of your small projects. Your FEMA PAC Crew Leader will schedule validation at a time convenient for you, and let you know which projects have been selected for validation.

- ▶ You are responsible for documenting all claimed costs. Be sure your project file includes all cost records, computations, measurements, notes, pictures, plans, Special Considerations, and any other documentation related to your project.

Large Projects

Large projects are funded based on actual costs. Here are the steps in processing a large project:

- Your FEMA Project Specialist and FEMA PAC Crew Leader work with you in preparing a Subgrant Application (*Project Worksheet*) describing damages, the scope of work to repair the damage, and a cost estimate.

- Funding is contingent on the resolution of any Special Considerations issues. See **APPENDICES B, C, D, AND E**.

- FEMA initially approves funding using the cost estimate and obligates the Federal share of the funds to the State.

- If the project's scope of work changes, even if it does not affect the total project cost, you need to notify both your FEMA Project Specialist and State PA Representative so that FEMA can screen the changes for potential environmental impact and ensure that the changes in the scope of work are eligible for Public Assistance Program funding.

- As the project proceeds, you may periodically request funds from the State to meet expenses that you have incurred or expect to occur in the near future. Expenses must be documented. In anticipating the need for payments to contractors, be sure to take into account the time that the State process requires for approval of requests and disbursement. Each State handles disbursements differently, and details for your State's procedures will be explained at the Applicants' Briefing.

- When the project is complete, you inform the State and the State will determine the final cost of accomplishing the eligible work, often performing inspections or audits to do so. The State then reconciles the actual costs with the Subgrant Application (*Project Worksheet*) estimate and transmits the information on the completed project to FEMA, certifying that the costs were incurred in the completion of eligible work.

- After reviewing the State's report and conducting such inspections or audits as necessary to verify eligible costs, FEMA may adjust (obligate/de-obligate) the amount of the Subgrant Application (*Project Worksheet*) to reflect the actual cost of the eligible work.

What do I need to do – large projects?

- Identify facilities, damage, and work needed.

- Take photographs of damage.

- You can expedite the handling of your large projects by identifying any Special Considerations (i.e., insurance, hazard mitigation, historic preservation, and environmental compliance, including floodplain management issues) that relate to your project early.

- The State cannot provide funds for costs that are outside the scope of work approved by FEMA. Therefore, you need to contact the State right away if you foresee or identify any changes to the scope of work or costs prior to or during performance of the work. Do not assume you can wait to report changes and that they will be approved. Since an inspection may be required, be sure that the damaged element can be inspected before it is repaired. Approval is needed to deviate from the scope of work and cost and a version to the Subgrant Application (*Project Worksheet*) is normally prepared.

- You are responsible for maintaining all source documentation and providing FEMA with the documentation needed to support your Subgrant Applications (*Project Worksheets*). All source documentation is needed for audit and closeout.

CHAPTER 7
Subgrant Applications (*Project Worksheets*) and Cost Estimating

General

Subgrant Applications (*Project Worksheets*) are the primary form used to document the information on your damages (i.e., locations, damage description and dimensions, scope of work, Special Considerations, and cost estimates). They supply FEMA with the information necessary to approve funding under the Public Assistance Program. Because the Subgrant Applications (*Project Worksheets*) are such important documents, FEMA will assign a FEMA Project Specialist to you to either help you complete the forms or complete the forms for you. The FEMA Project Specialist will call upon other Technical Specialists, as needed, to assist in developing the damage description, scope of work, and cost estimates, and to help you prepare hazard mitigation proposals, review insurance coverage, identify historic preservation issues, identify environmental issues, etc.

Each project must be documented on a separate Subgrant Application (*Project Worksheet*). Generally, each facility requires a separate Subgrant Application (*Project Worksheet*), but sometimes work for multiple facilities can be combined if there is a logical method of performing work required as a result of the declared event. This option offers flexibility in organizing and managing the work around your needs. For example, work that occurs at multiple sites, such as repairs to several washouts along a road or debris removal within specific boundaries could be shown as one project, as could work within a single jurisdiction (for example, all work in a park may be a project). Work also may be combined by:

- Type of facility (e.g., all sewer pump stations)
- System (e.g., repair of several breaks in a water distribution system, repair of paved roads, or repair of gravel roads)

- Boundaries (e.g., power lines in sectors or roads in a district)
- Method of work (e.g., work under one contract, work by force account labor, or a group of contracts under one contractor)

The FEMA PAC Crew Leader is responsible for ensuring that the resulting project is logical and consistent with other FEMA guidance.

There are some other things you need to know about combining work.

- Categories of permanent work (Categories C-G) may not be combined.
- A group of facilities at one location (e.g., a campus) may not be combined to be treated as a single facility or described as one project.
- Emergency work and permanent work may be combined into one project only when the emergency work is incidental to the permanent work. If the combination is appropriate, the emergency work needs to be evaluated separately from the permanent work to meet varying eligibility requirements (e.g., force account regular time is eligible for only permanent work).
- If multiple sites are combined into one project with a total estimated cost above the large project threshold, that project will be considered a large project.
- Work with an estimated or actual cost of less than $1,000 is not eligible. If there is a reasonable relationship to other sites, they may be combined to meet the minimum threshold. They are eligible only if they are reasonably related.
- Including a site with Special Considerations issues with other sites in one project may delay funding for all if compliance issues are identified. The sites with Special Considerations should be prepared on separate Subgrant Applications (*Project Worksheets*).
- Single facilities may not be broken into multiple projects to create small projects.

Once you have consolidated similar work items into projects, you will need to fully document your damage and repair plan by completing a Subgrant Application (*Project Worksheet*) for each project. Although more than one site can be combined to make a project, only one project may be listed on a Subgrant Application (*Project Worksheet*).

Your FEMA PAC Crew Leader will explain the advantages and disadvantages so that you can decide what works best for you.

Subgrant Application (*Project Worksheet*) Preparation

- **DISASTER.** Record the FEMA four-digit major disaster declaration number, disaster type, and the two-letter state abbreviation. For example:

    FEMA-1234-DR-AL

    FEMA-3305-EM-OK

- **PROJECT NUMBER.** For each Subgrant Application (*Project Worksheet*), two Project Numbers will be assigned: one defined by the FEMA Project Specialist during the development of the Subgrant Application (*Project Worksheet*) (and entered on the Subgrant Application (*Project Worksheet*) as it is developed) and an official one automatically assigned by the National Emergency Management Information System (NEMIS)/Emergency Management Mission Integrated Environment (EMMIE), FEMA's electronic tracking systems, when the Subgrant Application (*Project Worksheet*) is entered into the automated system. The Project Number assigned during Subgrant Application (*Project Worksheet*) development can reflect a FEMA-prescribed format, or if none, the applicant's own tracking system number. If neither is being used, the FEMA Project Specialist will assign a tracking number. The number may be no longer than seven characters and may include alpha, numeric, and special characters.

- **PUBLIC ASSISTANCE IDENTIFICATION NUMBER (PA ID NO).** Each applicant is assigned a unique identification number by FEMA. It is sometimes referred to as the Applicant's Federal Information Processing Standards (FIPS) Number or the Applicant's Identification Number. The number should be entered in the following format: XXX-XXXXX-XX. The first three digits identify the county where the applicant is located. If the first three digits are 000, the applicant is a State agency or has multiple facilities Statewide. The next five characters identify the particular applicant. The last two digits are used to identify departments or subdivisions within the applicant's agency or organization. If you wish to have subdivisions within your PA ID NO (e.g., Police Department, Public Works Department, etc.), discuss that possibility with your FEMA PAC Crew Leader.

- **DATE.** This is the date the Subgrant Application (*Project Worksheet*) was prepared. The format is MM/DD/YYYY.
- **CATEGORY.** The FEMA Project Specialist will designate the category of work in this block. The possible categories are:
    - Emergency   A - Debris Removal
    - Emergency   B - Emergency Protective Measures
    - Permanent   C - Roads and Bridges
    - Permanent   D - Water Control Facilities
    - Permanent   E - Buildings and Equipment
    - Permanent   F - Utilities
    - Permanent   G - Parks, Recreational Areas, and Other Facilities
- **DAMAGED FACILITY.** Identify the facility and describe its primary function.
    - If the project is a single site, enter the name of the facility and its basic function. For example, enter "Town Building C – Community Center", "Debris Removal", "Emergency Protective Measures", "County Road 66" or "City Memorial Hospital."
    - If the project consists of multiple sites, a general facility name can be provided in the block adding the reference "See Damage Description."
    - If the project involves Emergency Work (Category A or B), identify both the type and location of response. For example, enter "Debris Removal – City of Charles Sector A" or "Police, Fire, and Rescue – Charles County Courthouse."
    - If services are provided over time and multiple Subgrant Applications (*Project Worksheets*) are to be prepared for distinct durations, identify the duration in this block to clearly distinguish the project. For example, enter "Police Response – June 16-20, 2008."
- **WORK COMPLETE AS OF.** Indicate the date the work was assessed and the percentage of work completed on that date. This is the date the FEMA and/or State PA Representative visited or reviewed the site. It may be different from the date provided in the **DATE** block. Record the date in the following format MM/DD/YYYY. If any percentage of

work is complete, actual costs and documentation should be provided for the work complete. The **SCOPE OF WORK** and **PROJECT COSTS** blocks should separate the details for "Work Completed" and "Work to Be Completed." If the percentage of work completed does not match the percentage of costs incurred, explain why that is the case in the **SCOPE OF WORK** block.

- **APPLICANT.** This is the name of the government or legal entity to which the funds will be awarded. Applicant names should not be abbreviated.

- **COUNTY.** Enter the name of the county in which the damaged facility is located. If the damage is located in multiple eligible counties and the applicant wishes to combine the work, either confer with the State on State preferences or record "Multi-County" in this block and identify the specific counties in the **DAMAGE DESCRIPTION AND DIMENSIONS** and **SCOPE OF WORK** blocks.

- **LOCATION.** The exact location of the damaged facility must be described. This information must be specific enough to enable field personnel to easily locate the facility if a site visit is necessary. Some examples for location are: street address, intersection, and sector. Referring to activities taking place on a countywide, citywide, or jurisdiction-wide basis is the least preferred method and requires more specific supplemental information in the **DAMAGE DESCRIPTION AND DIMENSIONS** block. If there are multiple sites, enter "Multiple Sites (4)" and "see **DAMAGE DESCRIPTION AND DIMENSIONS**" (or other location at which they are entered on the Subgrant Application (*Project Worksheet*). Do not use facility names, as they may change over time.

- **LATITUDE AND LONGITUDE.** Enter the coordinates of the damaged site(s). If there are multiple sites, enter a primary location and identify the rest in the **DAMAGE DESCRIPTION AND DIMENSIONS** block. Do not use degrees, minutes, and seconds; instead, use up to three numbers before and up to five numbers after the decimal point.

- **PREPARED BY/TITLE/SIGNATURE.** Enter the name of the preparer and the person's position title. The hard copy must be signed.

- **APPLICANT REPRESENTATIVE/TITLE/SIGNATURE.** Enter the name and title of your representative. The signature will generally indicate your concurrence with the Subgrant Application (*Project

Worksheet) as prepared. If you do not concur with the Subgrant Application (*Project Worksheet*), describe the items of non-concurrence in the **SCOPE OF WORK** or in an attached memorandum or narrative. An applicant's signature is not required in order to process the Subgrant Application (*Project Worksheet*).

- **DAMAGE DESCRIPTION AND DIMENSIONS.** The description of disaster-related damage to the facility, including the cause of the damage and the area affected, is entered in this block. This block supports the basic eligibility determination for the work and defines the expectations for the scope of work and associated costs. The primary items that must be included in this section are:

  ▶ Description of the cause of damage. Provide a brief description of how the damage to the facility occurred, or what conditions of the disaster required the emergency service to be provided. For emergency work to be eligible, the Subgrant Application (*Project Worksheet*) must demonstrate that the disaster conditions caused an immediate threat. Therefore, describe the threat to health and safety or the threat to improved property.

  ▶ The party responsible for fixing the damage (with ownership documents, leases, etc.).

  ▶ Description of the pre-disaster condition of the facility. Include the pre-disaster design, function, and capacity and, if the facility was not in active use at the time of the disaster, explain why.

  ▶ Quantification of specific disaster-related damages or emergency services provided.

  ▶ The latitude and longitude of each site, if there are multiple sites.

  A Continuation Sheet can be used if more space is needed to continue the description. Make a statement at the end of the block, such as "See Continuation Sheet."

- **SCOPE OF WORK.** The description of work that has been completed and work that needs to be completed to repair disaster-related damages is entered in this block. The primary items that must be included are:

  ▶ Description of the work necessary to remove and dispose of disaster-related debris, conduct emergency response measures, or repair or replace the disaster-damaged facility to the pre-disaster condition.

- Identification of "Work Completed" and "Work to Be Completed."

- Descriptions of any Special Considerations (i.e., insurance, hazard mitigation, historic preservation, and environmental compliance issues, including floodplain management) that affect the scope of work.

- Description of any work that will restore the facility beyond its pre-disaster design, function, or capacity. Include relocations, replacements, alternate and improved projects, hazard mitigation proposals, and codes and standards.

- Description of the basis for the cost estimate. The project cost section of the Subgrant Application (*Project Worksheet*) is limited to cost data. Therefore, a description of how the costs were determined should be included here. It should identify how the work was or will be done (e.g., by force account labor), the methodology for determining the cost, and an explanation of why the costs are reasonable.

FEMA will also document ineligible work, associated costs, and the basis of ineligibility in this block.

- **SPECIAL CONSIDERATIONS.** Special Considerations is a term used by the Public Assistance Program to capture all program issues other than eligibility. For disasters, these are usually insurance, hazard mitigation, historic preservation, environmental compliance, and floodplain management. Four questions on the Subgrant Application (*Project Worksheet*) address these key issues. If the answer to any of the questions is "yes" or "unsure," an adequate explanation should be provided on the *Special Consideration Questions* form. FEMA Technical Specialists in the indicated areas will need to work with you on those issues.

- **PROJECT COST.** Project costs will be either estimated or actual. Only reasonable costs are eligible for FEMA assistance. A cost is reasonable if, in its nature and amount, it does not exceed that which would be incurred by a prudent person under the circumstances prevailing at the time the decision was made to incur the cost. In other words, a reasonable cost is a cost that is both fair and equitable for the type of work performed. The reasonable cost requirement applies to all labor, materials, equipment, and contract costs awarded for the performance of eligible work.

Getting your cost data. Once you have described your project, the eligible costs need to be determined. If the work has been completed, report the actual costs. If it has not been completed, estimate the costs. Generally, costs that are directly related to the performance of eligible work are eligible. The costs must be:

- Reasonable and necessary to accomplish the work
- Reduced by all applicable credits, such as anticipated insurance proceeds and salvage values
- Compliant with Federal, State, and local requirements for competitive procurement, if the work is done by contract

For small projects, estimates are especially important because they become the fixed amount of the grant. For large projects, estimates will be adjusted as reasonable actual costs become known, but their accuracy is important for your budgeting and management. FEMA makes the final determination on the reasonableness of a cost.

For large projects, make cost estimates early in the process before the design phase begins. However, a detailed design of the restoration work may need to be prepared for complex projects before a viable cost estimate can be developed. In such cases, a grant for engineering and design services may be approved first. Once the design is complete, a cost estimate for the work is prepared or actual bids for the work are used as the basis for the grant.

Costs for managing a project may also be included if the project is sufficiently large or complex to require them. Most small projects do not require project management above the first-level supervisor.

The most common methods of cost estimating are:

Unit costs – With this method, the project is broken down into elements based on the quantity of material that must be used to complete the work. For example, a culvert repair may be broken down into linear feet of pipe, cubic yards of fill, and square feet of pavement. The estimate for each of these items is a cost per unit that includes all labor, equipment, and material necessary to install that item (referred to as an in-place cost).

Unit cost data developed by State or local governments may be used for estimating costs, if appropriate. Alternatively, commercially available cost-estimating guides or data from local vendors and contractors may also be used. Also, FEMA has developed a list of unit costs for typical disaster repairs that may be used for estimating total costs. The order of preference for these cost data sources is: State and local data from previously completed projects, commercial estimating sources, and then FEMA cost codes.

<u>Time, equipment, and materials for local force account work</u> – This method may be used on projects that have been completed or will be completed by your employees, using your own (or rented) equipment and material purchased by you (or from your stock on hand). This method breaks costs down into labor, equipment, and materials. Costs must be thoroughly documented by payroll information, equipment logs or usage records, site or location, and other records, such as materials invoices, receipts, payment vouchers, warrants, or work orders. Final payment is based on documentation of your costs.

<u>Equipment rates</u> – A listing of equipment rates based on national data is available on the FEMA Web site at **www.fema.gov/government/grant/pa/eqrates.shtm**. These rates, or your established rates, whichever are lower, should be used to compute applicant-owned equipment rates. Applicant equipment rates approved under State guidelines are usually eligible up to $75 per hour.

Costs for use of cars and pick-up trucks are reimbursed on the basis of mileage if that rate is less costly than hourly rates. For all other types of equipment, rates typically include operation (including fuel), insurance, depreciation, and maintenance; they do not include the labor of the operator. As you complete your request, remember that labor hours need to match equipment use hours. Equipment used for less than half of the normally scheduled working date is reimbursable only for the hours used; however, if you can document that the equipment was used intermittently for more than half of the normally scheduled working hours for a given day, the entire day may be claimed. Stand-by time is not eligible.

FEMA rates do not apply to contracted or rental equipment, unless the equipment is rented from another public entity.

<u>Contracts</u> – Competitively bid contracts are used to summarize costs for work that you obtain from an outside source. In general, contract costs are for work already completed. If work has not yet begun on a project, but a contract has been bid or let for the eligible work, then the contract price can be used. To be eligible for reimbursement, procurements need to be awarded in compliance with State and

local procurement regulations and, generally, through full and open competition.

The Stafford Act requires communities to give preference to local firms in the awarding of contracts in major disasters and emergencies to the extent it is feasible and practicable. This includes contracts with local organizations, firms, and individuals that support response and recovery activities in a declared major disaster or emergency area, such as debris clearance, distribution of supplies, and reconstruction. Preference may be given through a local area set-aside or an evaluation criterion.

You must follow your established procurement policies when procuring goods or services with FEMA Public Assistance funds. Your established policies must adhere to all State, Tribal, and local laws and regulations as well as Federal standards. If your procurement standards do not comport with FEMA regulations, you must follow the procurement standards outlined in 44 CFR Part 13.36, **Procurement**.

You should take the necessary steps to ensure there are opportunities to award contracts to minority, women-owned, and Labor Surplus Area businesses and firms whenever possible. Examples of affirmative steps include, but are not limited to: ensuring that the group of contractors considered for awards reflects appropriate demographics, placing qualified minority and women-owned businesses on solicitation lists, soliciting these businesses when possible, and requiring prime contractors to take these steps as well.

FEMA will review, on an ongoing basis, procurement documents to validate that affirmative steps to utilize minority and women-owned businesses are being followed during the administration of an award. Individuals, organizations, and vendors can challenge contract awards through formal protest, complaints of discrimination in awarding the contract, allegations of real or apparent conflict of interest, or failure to follow established procedure for making an award. It is your responsibility to have protest procedures to handle and resolve disputes relating to your procurements and to provide information regarding protests to FEMA.

Procurement policies must include procedures to handle protests and disputes related to contracts awarded. You will be responsible for settling contractual and administrative issues involving your procurements, including protests, disputes, and claims. FEMA will not review any procurement protests or disputes unless the matter is a possible violation of Federal law or a violation of State, Tribal, and local laws protest procedures. Violations of State and local law will be referred to the appropriate jurisdiction. However, allegations of discrimination in the procurement and award of contracts will be referred to FEMA's Office of Equal Rights for processing.

If you fail to comply with the conditions of the award, including the provisions for procuring goods or services (in compliance with 44 CFR §§ 13.36, **Procurement**, and 13.37, **Subgrants**), FEMA may take enforcement actions outlined in 44 CFR § 13.43, **Enforcement**. These enforcement actions include, but are not limited to: temporarily withholding payment, disallowing all or part of a cost, wholly or partly suspending or terminating an award, or withholding future awards.

FEMA provides reimbursement for three types of contracts:

- Lump sum – For work within a prescribed boundary with a clearly defined scope and a total price.

- Unit price – For work done on an item-by-item basis with prices broken out per unit.

- Cost plus fixed fee – Either a lump sum or unit price contract with a fixed contractor fee added into the price.

Time and materials contracts should be avoided except when the work is necessary immediately and a clear scope of work cannot be developed quickly enough. Generally, time and materials contracts should be limited to no more than 70 hours. Time and materials contracts are acceptable for power restoration work. If time and materials contracts are used, you need to award the work competitively, monitor the work closely, and document contractor expenses carefully. A cost ceiling or "not-to-exceed" provision should also be included in the contract. For example, time and materials contracts for debris should be limited to a maximum of 70 hours of actual debris clearance work and should be used only after all available local, Tribal, and State government equipment has been committed. These contracts should be terminated once the designated dollar ceiling or not-to-exceed number of hours is reached. The contracts may be extended for a short period of time when absolutely necessary, for example, until unit price contracts have been prepared and executed.

Cost plus a percentage of cost and "piggyback" (expanding a previously awarded contract) contracts are not eligible. If you do not comply with the Federal procurement standards, FEMA may separately evaluate and reimburse only those costs it finds fair and reasonable

Using Your Cost Data in the Subgrant Application (*Project Worksheet*). For small projects (at least $1,000 and less than $63,200 in fiscal year 2010), you will estimate costs using the cost data developed through the methods outlined above.

The method for estimating for large projects (equal to or more than $63,200 in fiscal year 2010) depends on the stage of completion when the *Project Worksheet* is prepared. If the project is more than 90 percent complete, use actual data for the completed work and estimate the remaining work using the methods outlined above. If the project is 90 percent or less complete, cost estimating will be a team effort with FEMA. Because final funding for large projects is based on actual costs to complete the eligible scope of work, your funding for each large project will be adjusted after all work is complete.

FEMA cost estimators and construction engineers use a cost estimating methodology called the Cost Estimated Format (CEF) to estimate your total costs for each project. The CEF is a forward pricing model based on construction industry standards that accounts for all possible costs associated with your project. The CEF provides better up-front estimates, allows you to budget project costs with greater confidence, and provides information that will help you manage your projects.

The CEF requires a clear definition of the scope of work eligible for public assistance. Once this scope of work has been developed, the CEF is applied in parts. Part A represents the base cost of completing the project. It includes the labor, materials, and equipment necessary to complete each item of the scope of work. Parts B through H contain job-specific factors that are usually added to the base cost determined in Part A.

<u>Part B</u> includes construction costs not typically itemized in Part A, such as the general contractor's supervision costs.

<u>Part C</u> includes construction cost contingencies and addresses budgetary risks associated with project complexity during the design process.

<u>Part D</u> accounts for the contractor's overhead, insurance, bonds, and profit.

<u>Part E</u> accounts for cost escalation over the life of the project.

<u>Part F</u> includes fees for special reviews, plan checks, and permits.

Part G is reserved for change orders, hidden damages, and differing site conditions discovered after construction starts.

Part H accounts for your cost to manage the design and construction of the project.

What do I need to do to help develop the Subgrant Applications (Project Worksheets)?

- You have 60 days following the first substantive meeting with FEMA (usually the Kickoff Meeting) to identify and report damaged facilities to FEMA.

- FEMA will work with you on describing your damages and scope of work, especially on large projects. But you need to ensure that all of your damages are identified and that your damages and scopes of work are well described. The costs are estimated based on this information. It is in your interest for the cost estimate to be as inclusive as possible.

- Make sure Special Considerations (i.e., insurance, hazard mitigation, historic preservation issues, and environmental compliance, including floodplain management) are identified.

- Contact your State to ensure you are using proper contracting guidelines.

- Ensure any procurement complies with Federal, State, and local requirements.

- Ensure your repairs are in accordance with the provisions of your final approved Subgrant Application (*Project Worksheet*).

- Monitor and document work, whether done by your own force account or contractors. If changes in the Subgrant Applications (*Project Worksheet*) are needed, request approvals right away.

- Ensure your cost records are organized and complete and include all required documentation.

- Working with the State and FEMA may be necessary to research and evaluate requirements for facilities with historic or environmental issues. If applicable, keep a record of the information pertaining to the alternatives that are considered.

Forms

FEMA Form 90-91 *Project Worksheet* (Subgrant Application)

FEMA Form 90-91A *Project Worksheet - Damage Description and Scope of Work Continuation Sheet* (Subgrant Application)

FEMA Form 90-120 *Special Consideration Questions*

(Forms are available at **www.fema.gov/government/grant/pa/forms.shtm**.)

Versions/Amendments/Change Orders

Original Subgrant Applications (*Project Worksheets*) are written when damages are identified. FEMA may generate subsequent versions/amendments/change orders for large projects to modify a scope of work, add damaged elements, grant time extensions, or modify the cost. Subsequent versions/amendments/change orders for small projects are rare and are generated only if the scope of work of the small project has changed very substantially, for example, due to a gross error or omission in the scope of work.

CHAPTER 8

Dealing With Changes to the Project and Appealing Decisions

Time

Eligible work must be completed within timeframes established by regulation. These timelines begin on the declaration date of the major disaster or emergency. The timelines correspond to type of work:

Debris Clearance	6 months
Emergency Protective Measures	6 months
Permanent Work	18 months

If you can justify a need for additional time due to extenuating circumstances or unusual project requirements, your State may grant a time extension for your work as long as there is no change in the scope of work or increase in cost. For debris removal (Category A) and emergency protective measures (Category B), the State may grant up to an additional 6 months (for a total of 12 months) for the completion of the scope of work. For permanent work (Categories C–G), the State may grant up to an additional 30 months (for a total of 48 months). If your detailed justification merits even more time, the FEMA Regional Administrator can approve it. (Categories A–G are discussed in **CHAPTER 3**.)

Because FEMA provides assistance only for those costs incurred up to the latest approved completion date for a particular project, completing your work within an approved timeline is important.

If you do not complete the approved scope of work, FEMA must rescind your funding for the project.

Cost and Scope

During the course of the approved work, you may find that work or costs are exceeding the approved project description and estimates. If you need to change the scope or cost of work on the Subgrant Application *(Project*

Worksheet), you must justify and request that change from the State. If the State agrees that the requested change is justified, the State will ask FEMA to approve the change. Any change in your project that affects the scope or cost of work can only be approved by FEMA.

Project changes are generally not approved on the basis of a cost overrun on an individual small project unless the overrun is very substantial, for example, due to a gross error or omission in the scope of work.

If you have a significant net overrun for all your small projects, you may appeal for additional funding. However, you should be prepared for a complete audit of all your small projects in response to such a request.

Potential deviations from the cost and scope of work described in the Subgrant Application (*Project Worksheet*) must be reported as soon as you are aware of them.

Undiscovered and Newly Discovered Damage

Assistance for any newly discovered damaged facilities must be requested within 60 days of the first substantive meeting (typically the Kickoff Meeting) with FEMA.

If you discover additional damage in a facility for which the damage description and scope of work have already been identified, report it immediately to the State. Don't assume that you can obtain additional funding at the end of the project. The State and your FEMA PAC Crew Leader will want to see the damaged element before it is repaired. FEMA will determine whether to approve or deny your request for additional funding.

Appeals

The appeals process allows you to request reconsideration of decisions regarding the provision of assistance. There are two levels of appeal, both of which are processed through your State to FEMA. The first level appeal is to the FEMA Regional Administrator. The second level appeal is to FEMA Headquarters. You must file an appeal with the State within 60 days of receipt of a notice of the action that is being appealed, and you must provide justification to support the appeal. The supporting material should include any relevant documents, for example, correspondence, contracts, bids, and insurance policies. Explain why you believe the original determination is wrong and the amount of adjustment being requested.

What do I need to do?

- Immediately advise your State PA Representative if you anticipate that your projects are falling behind schedule or becoming inconsistent in any way with the scope of work or costs described in your Subgrant Application (*Project Worksheet*).

- Promptly appeal any FEMA decision with which you disagree through your State PA Representative.

CHAPTER 9
Documentation

You are responsible for establishing and maintaining accurate records of events and expenditures related to disaster recovery work for which you request FEMA assistance. The documentation required describes the "who, what, when, where, why, and how much" for each item of cost.

These records become the basis for recovering your final project costs, and, for small projects, will be used to sample and validate your estimated project costs. Most States require you to submit documentation on costs before they will disburse any funds to you for your large projects. Proper documentation will help you justify cost overruns or time extensions. You also need cost information to develop your projects, show alternatives considered when an environmental or historic assessment was required, show that there has been no duplication of benefits, and prepare for audits or other program reviews. **Incomplete or improper records can lead to either partial or total loss of assistance**.

There are many ways to maintain your records. FEMA provides some forms that will help you. However, you may use your own records systems to replace the FEMA forms and to record other information. What is important is that you have the necessary information tied to each specific project readily available and in a usable format.

Because work may be done before a Subgrant Application (*Project Worksheet*) is written, you should immediately establish a file for each site where you identify damage. This may involve opening a file for each building, vehicle, location (e.g., of road washout damage), and type of work (e.g., police response).

Depending on the type of damage, you may need to keep:

- Records that demonstrate the presence of an immediate threat
- Drawings, sketches, and plans of pre-disaster facility design (to scale)
- Drawings and sketches of disaster-related damages (to scale)
- Drawings and sketches of completed or proposed repair (to scale)
- Calculation sheets detailing specific dimensions and quantities of damage

- Force account labor records (i.e., payroll information, timesheets, and administrative policies)
- Temporary hire labor records (i.e., work for which the labor was needed, payroll information, and timesheets)
- Fringe benefit calculations
- Force account equipment usage information and rate schedules
- Records of materials from inventory
- Rental and lease agreements
- Photographs of site, overall facility, specific damage, and repairs
- Subgrant Applications (*Project Worksheets*)
- Site location maps
- Flood Insurance Rate Maps
- Facility maintenance records (e.g., for roads or debris basins)
- Facility inspection/safety reports (as may be available for bridges and dams)
- Engineering/technical reports and specifications for repair
- Codes and standards governing repairs/replacements
- Insurance information (i.e., policies, proof of loss statements from insurance company, deductible information, etc.)
- Documents supporting compliance with environmental and historical preservation issues
- Hazard mitigation proposals (as allowed under Section 406 of the Stafford Act)
- Justification for requests for a relocation, improved, or alternate project
- Records of donated labor, materials, and equipment, including location, description of work, name of worker, hours worked, value per hour, and certification
- Contracts or contractor bids (including invoices and copies of payments)

- Inspection logs for work included in Subgrant Applications (*Project Worksheets*)
- Permits
- Correspondence
- Invoices/warrants/checks
- Job orders
- Mutual aid agreements and records of mutual aid requests and receipt

What do I need to do?

▶ Maintain accurate disbursement and accounting records to document the work performed and the costs incurred. You are responsible for substantiating all costs and your records must be complete and organized.

▶ Designate a person to coordinate the accumulation of records. Some applicants also use the services of an internal auditor.

▶ Establish a file for each project where work has been or will be performed. For projects that include more than one physical site, maintain records showing specific costs and scopes of work by site.

▶ File all of the documentation pertaining to a project with the corresponding Subgrant Application (*Project Worksheet*) and maintain the files as the permanent record of the project. Ensure your documentation can be retrieved by the project number assigned on your Subgrant Application (*Project Worksheet*).

▶ Examine the FEMA forms to see if they will be of use to you in recording your information.

- Keep all financial and program documentation for three years beyond the date of the State's Final Status Report (FSR), or longer if required by the State's record retention policies. If an FSR is not required (find out from your State PA Representative), records must be maintained for three years from the date of the final certification of completion of your last project or as required by the State. Records are subject to audit by State auditors, FEMA, and the U.S. Department of Homeland Security's Office of Inspector General.

FORMS

FEMA Form 90-123 *Force Account Labor Summary Record*

FEMA Form 90-127 *Force Account Equipment Summary Record*

FEMA Form 90-124 *Materials Summary Record*

FEMA Form 90-125 *Rented Equipment Summary Record*

FEMA Form 90-126 *Contract Work Summary Record*

FEMA Form 90-128 *Applicant's Benefits Calculation Worksheet*

(Forms are available at **www.fema.gov/government/grant/pa/forms.shtm**.)

CHAPTER 10
Progress Reports, Closeout, and Audit

Progress Reports

FEMA expects you to actively manage your approved work, setting targets and making sure they are met. FEMA requires the State to report on your projects. Therefore, the State will require from you progress reports on each of your projects. Progress reports are a critical part of disaster management. They ensure that FEMA and the State have up-to-date information on the work and expenditures. Reporting requirements generally focus on large projects and include information such as: status of each project, projected completion dates, amount of expenditures for each project, and any circumstances that could delay the project or result in noncompliance with the conditions of the FEMA approval. Your State will inform you of the reporting requirements for the disaster.

Closeout

The purpose of closeout is to certify that all work has been completed and all of your eligible costs have been reimbursed. It is an important last step in the process.

Upon completion of a large project, you must submit documentation to the State to account for all incurred costs. The State is responsible for ensuring that all incurred costs are associated with the approved scope of work and for certifying that work has been completed in accordance with FEMA standards and policies. The State then submits documentation of project costs to FEMA for review. FEMA may conduct a final inspection of the site as part of this review. Once the review is complete, FEMA determines whether additional funds should be obligated or whether funds should be de-obligated for the project.

Once your small projects are complete, the State must certify that all work has been completed in accordance with the approved scope of work and in compliance with FEMA standards and policies, and that all payments have been made. This certification does not specify the **amount** spent on the projects, only that the projects were completed. Should you fail to complete any small project, including any mitigation work, funds approved for the project will be de-obligated. If you spend less than the amount approved by FEMA, no adjustment is made. However, if you incur costs significantly greater than the total amount approved for all your small projects, you may appeal for additional funding. Remember, this opportunity applies only to a significant net cost overrun for **all** small projects combined, not to an overrun for an individual small project. You should be prepared for a complete audit of all your small projects if you appeal for additional funding.

Closeout procedures are different for each State; however, you should notify the State PA Representative immediately as you complete each large project and when all of your small projects have been completed. You do not need to await the completion of your large projects to proceed with closeout of your small projects.

Audits

All documents are subject to audit by the State, FEMA, and the U.S. Department of Homeland Security Office of Inspector General. Because failure to properly document any claimed expenses may result in loss of funding, working within your approved scope of work and costs and documenting each project thoroughly are critical.

What do I need to do?

- ▶ Keep an eye on the calendar. From the date of the Presidential declaration, you have 6 months to complete emergency work and 18 months to complete permanent work. You should complete your work within those periods, but you may obtain extensions from the State based on extenuating circumstances or unusual project requirements beyond your control. You must complete your project within an approved timeline or risk loss of funds.

- ▶ Notify the State immediately as you complete each large project. Large projects are closed out individually as each is completed. Costs must be reconciled (difference between estimated and actual costs for eligible work) as each large project is completed. You must account for all incurred costs. If you have a cost overrun, the quality of your documentation and justification will be critical to obtaining additional funds.

- ▶ Notify the State immediately when you have completed your last small project. Small projects are closed out as a group.

- Certify to the State that all funds were expended and the project scope of work is complete.

- You are responsible for documenting all costs associated with your projects. Make sure all documentation for a project is accurate, complete, and up to date for closeout review. Failure to properly document a project may result in loss of funding for any claimed work.

- Maintain program and financial records for three years beyond the date of the State's final FSR for the grant or longer if required by the State's record retention policy. If an FSR is not required, records must be maintained for three years from the date of the final certification of completion of your last project. Check with your State for the record retention requirements that pertain to you.

CHAPTER 11
Conclusion

Sources of Information

This Handbook was written to give you basic information to obtain assistance under the Stafford Act. Your State PA Representative and FEMA PAC Crew Leader and their staff will help you with your recovery. Further information about the Public Assistance Program is available on the FEMA Web site **www.fema.gov/government/grant/pa/index.shtm**. In addition to an electronic version of this Handbook, the *Public Assistance Guide* (FEMA 322), which includes the Stafford Act and relevant portions of Title 44 of the Code of Federal Regulations, the *Public Assistance Policy Digest* (FEMA 321), the *Public Assistance Debris Management Guide* (FEMA 325), specific FEMA disaster Assistance policies and fact sheets, CEF tools, and other forms and tools that will help you understand the benefits to which you are entitled are available on the site.

Apply for Assistance

The FEMA Public Assistance Program can help you if:

- You are an eligible applicant.
- You are legally responsible for an eligible service or for an eligible facility damaged in the declared event.
- The work and reasonable costs meet FEMA requirements.

Keep Good Records

- Establish, organize, maintain, and retain documentation.

Your Checklist of Milestone Events

- Conduct a local assessment and report damages to the State.
- Participate in Preliminary Damage Assessment, if you have an opportunity.
- Watch for declaration of emergency by the Governor.

- Watch for declaration of a major disaster or emergency by the President.
- Designate a representative to interact with the State and FEMA.
- Attend the Applicants' Briefing scheduled by the State.
- Submit your Pre-application (*Request for Public Assistance*) form to the State.
- Attend the Kickoff Meeting scheduled for you, the State, and FEMA.
- If needed, request Immediate Needs Funding or Expedited Payments.
- Identify damages within 60 days of your Kickoff Meeting.
- If you prepare them yourself, submit Subgrant Applications (*Project Worksheets*) for small projects within 60 days of your Kickoff Meeting.
- Complete Subgrant Applications (*Project Worksheets*) with the FEMA Project Specialist and Technical Specialists, if appropriate, to identify scope of work, costs, and Special Considerations (i.e., insurance, hazard mitigation, historic preservation, and environmental compliance including floodplain management) pertaining to your damages and recovery plans.
- Be aware of the timelines for accomplishing work. From the date of Presidential declaration:
  - 6 months for debris removal and emergency protective measures
  - 18 months for permanent work
  - If there is no change in scope of work or increase in costs, the State may approve extensions for up to 12 months for debris removal and emergency protective measures and up to 48 months for permanent work. Other extensions require FEMA approval.
- Be mindful that temporary relocations of essential facilities are limited to 6 months and that extensions require justification.
- Manage the accomplishment of work and costs.
- Participate in site inspections by the State and FEMA.

- Cooperate with progress reporting as required by the State.
- File appeals if you disagree with an action or decision:
    - Within 60 days of receipt of notice of the action or decision being appealed
    - For net small project overruns, within 60 days of the completion of all small projects
- Close out small projects when all are completed.
- Close out large projects as each project is completed.
- Cooperate with audits of your program and financial records by the State, FEMA, and the U.S. Department of Homeland Security Office of Inspector General at any time during the process.

APPENDIX A
Private Nonprofit (PNP) Organizations

PNP organizations that own or operate facilities that provide certain services of a governmental nature are eligible for assistance. However, the organizations, facilities, and services must meet some additional eligibility criteria beyond those that apply to governmental applicants.

Qualifying PNP Facilities

Qualifying PNP facilities are:

- Educational – These are primary, secondary, and higher education facilities, including vocational facilities. Unless used primarily for religious purposes, eligibility extends to buildings, housing, and classrooms, plus related supplies, equipment, machinery, and utilities necessary for instructional, administrative, and support purposes.

- Utility – This includes buildings, structures, and systems, even if not contiguous, of energy, communication, water supply, sewage collection and treatment, or other similar public service facilities.

- Emergency – These include buildings, structures, equipment, or systems used to provide emergency services, such as fire protection, ambulances, and rescue, even if the facilities are not contiguous.

- Medical – These include hospitals, clinics, outpatient services, hospices, nursing homes, and rehabilitation centers that provide medical care. Eligible components include the administrative and support facilities essential to the operation of the facility, even if not contiguous.

- Custodial care, including facilities for the aged and disabled – These include those buildings, structures, or systems, including those essential for administration and support, that are used to provide institutional care for persons who do not require day-to-day medical care, but do require close supervision and some physical constraints on their daily activities for their protection.

- Certain irrigation facilities – These include irrigation facilities that provide water for essential services of a governmental nature. Eligible irrigation facilities include those that provide water for fire suppression, generating electricity, and drinking supply. Facilities that provide water for agricultural purposes are not eligible. For an irrigation facility element with mixed purposes, only damages related to the eligible purpose are eligible.
- Facilities on Indian reservations.
- Other essential governmental services, including:
  - Museums
  - Zoos
  - Performing arts facilities
  - Community arts centers
  - Community centers
  - Libraries
  - Homeless shelters
  - Senior citizen centers
  - Rehabilitation facilities
  - Shelter workshops
  - Health and safety services of a governmental nature, such as low income housing, alcohol and drug treatment centers, facilities offering programs for battered spouses, facilities offering food programs for the needy, and daycare centers for children and those with special needs

If you operate such a facility or service, you may be eligible if you also have:

- An effective ruling letter from the U.S. Internal Revenue Service granting tax exemption under Section 501(c), (d), or (e) of the Internal Revenue Code of 1954, as amended, or
- State certification that your organization is a non-revenue producing nonprofit entity organized or doing business under State law.

Facilities must to be open to the general public. However, PNPs on Indian reservations and some educational, utility, emergency, medical, and custodial care services are exempt from this requirement.

Qualifying Work

For eligible PNPs performing eligible functions, debris removal from the facility's property, emergency protective measures to prevent damage to an eligible facility and its contents, and repair or replacement of eligible damaged facilities are eligible for assistance. The potentially eligible facility is the one from which the qualifying service is delivered.

- All eligible PNPs seeking reimbursement from FEMA for debris removal and emergency protective measures should apply directly to FEMA for assistance.

- Eligible PNPs seeking assistance with permanent repairs and restoration apply for disaster assistance according to the following requirements, depending on whether the facility provides critical or non-critical services.

    PNP organizations that supply critical services should apply directly to FEMA for cost-shared assistance. Facilities with a mix of critical and non-critical services should also apply directly to FEMA. Critical services are defined as:

    - Power – Generation, transmission, and electrical power distribution facilities

    - Water – Treatment, transmission, and distribution facilities for a water company supplying municipal water (also, water provided by an irrigation company for potable, fire protection or electricity generation purposes)

    - Sewer and wastewater – wastewater collection, transmission, and treatment facilities

    - Communications – Telecommunications transmission, switching, and distribution facilities

    - Education – Primary, secondary, and higher education facilities (including vocational facilities)

    - Emergency medical care – Hospitals, clinics, outpatient services, hospices, nursing homes, and rehabilitation centers that provide medical care

> ▸ Fire protection/emergency services – Fire and rescue companies, including buildings and vehicles essential to providing emergency services and ambulance companies

Non-critical services are those that do not qualify as critical service facilities. PNPs with non-critical services should apply to FEMA first, but also must apply to the Small Business Administration (SBA) for a low-interest loan for repair of disaster damages.

- Facilities that support PNP facilities providing critical services (e.g., administration buildings and parking garages) are not eligible for the critical designation.

- Facilities that provide critical services should apply directly to FEMA for cost-shared assistance. Facilities with a mix of critical and non-critical services should also apply directly to FEMA.

Small Business Administration

Eligible PNP facilities that do not provide critical services must apply for a loan from the SBA for **permanent** work. If declined for a loan, the PNP may be eligible for FEMA assistance. Also, if the loan does not cover all eligible damages, FEMA may be able to provide cost-shared assistance for the difference. FEMA cannot take action on your request for funding for permanent work while your loan request is pending with the SBA, but you still should identify your damages to FEMA immediately so that Subgrant Applications (*Project Worksheets*) can be prepared to record damages, and to begin the process for funding your **emergency work** (which does not require an SBA loan application). Having the Subgrant Application (*Project Worksheet*) for your permanent work on file will allow FEMA to expedite your assistance once the SBA renders its decision on your loan request.

Some Things You Need to Know

Facilities with a mix of eligible and ineligible activities (e.g., a clinic that has a commercial pharmacy in the building) are eligible if over 50 percent of the space or time is used for eligible activities. Benefits will be pro-rated. If less than 50 percent of the space or time is used for eligible activities, the facility is not eligible at all.

Operating costs for providing services are not eligible, even if they are increased by the event. Labor, material, and equipment costs for providing assistance to disaster victims are also ineligible, even if the services are outside the organization's basic mission. If you provide services under contract to a State agency or local government agency, the costs may be eligible if they are claimed by the agency or government. Some reasonable short-term additional costs that are directly related to accomplishing specific approved emergency health and safety tasks as part of eligible emergency protective measures may be eligible if they can be documented and connected to a specific emergency task (e.g., increased utility costs for a permanently mounted generator at a hospital).

Homeowners' associations and gated communities are not eligible for Public Assistance funding, except when:

- Removal of debris from roadways within the community is required to create an emergency path of travel. The work is performed or contracted by an eligible local or State-level government entity with legal authority and reimbursement is requested by the eligible local or State entity.
- The association meets the criteria for an eligible PNP that provides eligible critical services (i.e., educational, medical, custodial care, emergency, and utility [but not irrigation] facilities/services).

Community Development Districts (CDD) are special districts authorized under State law to finance, plan, establish, acquire, construct/reconstruct, extend/enlarge, equip, operate, and maintain systems, facilities, and basic infrastructure within their respective jurisdictions. To be eligible, the CDD must be established under State law, have legal responsibility for ownership, maintenance, and operation of an eligible facility, and be open to the general public. If access is restricted, a CDD is only eligible when:

- Debris removal from roadways is required to create an emergency path of travel, the work is performed or contracted by an eligible local or State-level government entity with legal authority, and reimbursement is requested by the eligible local or State entity. If the CDD was established for the purpose of road maintenance, the CDD may do the work and apply as an applicant.
- Emergency protective measures required for repair of facilities for which the CDD was created. If it was created for water and sewer operations, the CDD may claim assistance for only those facilities. Other facilities of the CDD would be ineligible for both emergency and permanent work.

Some of the PNP facilities that are not eligible for FEMA assistance include:

- Recreational facilities
- Job counseling or job training facilities
- Conference facilities
- Facilities for advocacy groups not directly providing health services
- Facilities for advocacy or lobbying
- Irrigation for agricultural purposes
- Roads owned and operated by a homeowners' association
- Parking facilities not directly supporting an eligible facility
- Facilities for religious service or education

What do I need to do?

- Identify damages.
- Establish your status as an eligible PNP organization.
- Apply for FEMA assistance.
- Apply for an SBA loan, if your work is a permanent repair for a non-critical service.

APPENDIX B
Insurance

Insurance is an important consideration of the Public Assistance Program and comes into play in two respects.

- FEMA cannot provide disaster assistance (for either emergency work or permanent work) if damages or losses are covered by insurance. Consequently, you need to file your claims with your carrier as soon as possible after the disaster.

- FEMA requires you to obtain and maintain insurance on facilities repaired/replaced with FEMA funding.

Insurance Reductions

Duplication of Benefits. FEMA must reduce all project grants for insured property by the amount of actual insurance proceeds received or by the amount of proceeds that can be reasonably anticipated from a review of the insurance policy. By taking this reduction, FEMA eliminates the potential for duplication of benefits for the same loss. This applies to both general property insurance and flood insurance. Insurance often covers buildings, contents of buildings, vehicles, equipment, debris removal, demolition, cleanup, snow removal, temporary facilities, code and standard upgrades, operating costs, and other losses. Reductions apply to both emergency work and permanent work under the Public Assistance Program.

For general property insurance (a term FEMA uses to describe all perils except for flood), FEMA will apply a reduction based on the statement of loss, if received, to reduce the eligible amount of funding by the amount of the actual insurance proceeds. However, if the insurance proceeds are unknown or do not appear to agree with the policy coverage description, a FEMA Technical Specialist in the insurance field will review your insurance policy with your Subgrant Applications (*Project Worksheets*) to determine this amount. The anticipated proceeds are then deducted from the original eligible amount of disaster damage.

For flood damages, the reduction of eligible costs is based on whether or not the damage is located within a Special Flood Hazard Area (SFHA) as defined by the NFIP. If the facility is outside the SFHA, the reduction is the actual or anticipated insurance proceeds. If the facility is within the SFHA and has no or insufficient flood insurance, eligible costs will be reduced by the lesser of:

- The value of the facility, or
- The maximum amount of insurance proceeds that would have been payable had the facility by covered through NFIP coverage.

It is the sole responsibility of the Public Assistance applicant to request a Letter of Map Amendment or a Letter of Map Revision if the applicant believes that a structure is not actually located in the identified SFHA as indicated on current Flood Insurance Rate Maps.

Past Damages. Project grants are also reduced if there was a requirement that insurance be obtained and maintained for the facility for the same peril.

Required Documentation

Documents. You must submit copies of all insurance documentation (including the insurance policy with all data, declarations, endorsements, exclusions, schedules and other attachments or amendments) to your FEMA PAC Crew Leader. Any settlement documentation (including copies of the claim, proof of loss, statement of loss, and any other documentation describing the covered items and insurance proceeds available for those items) must also be submitted. This documentation will be used to determine your level of project funding. Insurance reductions will be made prior to project approval and noted in the General Comments section of the Subgrant Application (*Project Worksheet*).

Apportionment. If your insurance policy covers both insured eligible and insured ineligible damages, such as both property and business income losses, without specifying limits for each type of loss, FEMA will apportion the anticipated recovery based on the ratio of insured eligible to insured ineligible damages.

Perils. The type of peril (e.g., flood, wind, fire, hail, etc.) that caused the disaster damage must be identified in the Subgrant Application (*Project Worksheet*). Insurance coverage often excludes certain perils and may only cover certain damaged items within a project. In addition, a single facility may have been damaged by multiple perils, such as wind and flood damage during a hurricane, and the insurance may only cover some of the perils. Damages from wind and flood must be separated on the Subgrant Application (*Project Worksheet*), as they are assessed very differently for insurance purposes.

Past Damages. Finally, you must report whether the facility has ever received disaster assistance from FEMA. If any required insurance coverage for the same peril was not obtained or maintained, FEMA may not provide assistance for the facility.

Frequent Claims. Uninsured losses that might be eligible for FEMA assistance include:

- Reasonable deductible in your first claimed FEMA assistance if the cost is accrued to you
- Depreciation (that is, the differences between FEMA eligible costs and final loss valuations used by insurers)
- Costs in excess of insurance policy limits, including sub-limits for certain perils (such as flood or earthquake)

Obtaining and Maintaining Insurance

Requirement. As a condition of receiving Federal assistance, you must obtain and maintain insurance in at least the amount of the eligible damage to protect against future loss to such property from the same peril for the useful life of the repairs. The required insurance coverage must be obtained, or letter of commitment accepted by the State, prior to the release of any Federal funds. If the insurance is not maintained, the facility will receive no assistance in future events for the same peril.

Insurance can be purchased for a variety of valuable properties. Generally, the following are insurable: buildings, contents of buildings, vehicles, and equipment. If any other specific type of insurance is reasonably available, adequate, and necessary to insure your facility, you may be required to obtain and maintain that insurance coverage.

If assistance was received for flooding, you are required to obtain flood insurance even if you are located outside the floodplain.

Amount of Insurance. You must obtain insurance for the full amount of FEMA assistance (including Section 406 hazard mitigation assistance), even if the amount is over the NFIP limit. If the amount of assistance is over the NFIP limit, you will need to obtain commercial insurance.

When Insurance Is Not Required. Insurance is not required if the estimate for the repair of disaster-related damage is less than $5,000. Insurance is also not required on temporary facilities.

Premiums Are Your Costs. Premiums (for temporary as well as for other facilities) are not eligible for reimbursement. However, prudent risk management practices encourage appropriate coverage for peril exposure at a facility. If a FEMA-funded temporary facility is damaged in a non-Presidentially declared event, FEMA will not repair or replace it.

What do I need to do?

- Contact your insurance carrier, file your claims, and pursue payments you are entitled to under your insurance policies.

- Collect your insurance coverage and settlement documentation.

- Discuss insurance coverage with your FEMA PAC Crew Leader or FEMA insurance Technical Specialist, and provide him/her with insurance documents and statements of loss. If the damaged facility is rented, a copy of the lease or rental agreement may also be necessary.

- Identify all past disaster damages and claims for Federal assistance, and discuss them with your FEMA PAC Crew Leader or FEMA insurance Technical Specialist.

- Be sure your damages and scope of work are adequately described in the Subgrant Application (*Project Worksheet*), even if a portion will be covered by your insurance. This will allow for a future adjustment of eligible costs if there is a change in your insurance settlement.

- Keep the FEMA PAC Crew Leader informed of any problems.

- Follow FEMA procedures if you believe your facility was incorrectly included in an identified SFHA and be sure your Subgrant Application (*Project Worksheet*) notes the situation and steps you are taking to correct the elevation data.

- Obtain and maintain insurance on FEMA funded damages as required. Obtain insurance coverage as soon as possible.

APPENDIX C
Hazard Mitigation

What It Is

If your damaged facility is eligible for permanent repairs, you may also be eligible for additional cost-shared assistance under Section 406 of the Stafford Act for cost-effective measures that will prevent future similar damage to your facility. These measures are called hazard mitigation measures. FEMA strongly encourages you to consider hazard mitigation opportunities as a part of the repair and restoration of your facility. Hazard mitigation measures for your project may be proposed by you, FEMA, or the State. While your basic funding will return your facilities to their pre-disaster design, hazard mitigation measures will improve on the pre-disaster design. (Upgrades required to meet applicable codes and standards are part of your basic eligible restoration work, not hazard mitigation measures.)

Hazard mitigation opportunities usually present themselves at sites where damages are repetitive and simple measures will solve the problem. A hazard mitigation proposal is a written description and cost estimate of what it will take to repair the damage in such a way as to prevent it from happening again. The proposal is submitted with the Subgrant Application (*Project Worksheet*) and describes in detail the additional work and cost associated with the mitigation measure. Hazard mitigation measures must meet one of the following tests of cost-effectiveness:

- Cost no more than 15 percent of the total eligible cost of eligible repair work for the damaged facility
- Cost no more that 100 percent of the total eligible cost of eligible repair work and on the list of FEMA-approved mitigation measures
- Have a benefit-cost ratio of equal to or greater than 1.0

Because mitigation measures can be technically complex and must be thoroughly evaluated for feasibility, you may want to ask your FEMA PAC Crew Leader for technical assistance in identifying hazard mitigation measures or in preparing a proposal. The FEMA PAC Crew Leader may request additional help from a Technical Specialist who can identify hazard mitigation opportunities and analyze hazard mitigation proposals for cost-effectiveness and feasibility. Since hazard mitigation will often change the pre-disaster design of the facility and will require consideration of environmental and historic preservation issues, your FEMA PAC Crew Leader may also obtain assistance from Technical Specialists in those areas.

If approved (i.e., good engineering design, cost-effective, eligible, technically feasible, and compliant with environmental and historic preservation rules, etc.), the additional work will be addressed in a separate paragraph within the **SCOPE OF WORK** section of the Subgrant Application (*Project Worksheet*). If approved, the costs will be included in the overall funding of your project.

Some Things You Need to Know

If your hazard mitigation proposal is approved by FEMA, you are required to complete the work while completing the repair documented on the Subgrant Application (*Project Worksheet*).

The mitigation measure becomes part of the work on which you are required to obtain and maintain insurance.

No measure may increase risks or cause adverse effects to the facility or to other property. For example, if the mitigation dealt with drainage issues, an important concern would be the effect the mitigation measure would have downstream.

Alternate projects, and improved projects if a replacement facility is involved, are not eligible for Section 406 (Stafford Act) hazard mitigation funding. Also mitigation funding that might have been approved for an original location does not transfer to a new location for an improved project that involves combining facilities.

Undamaged elements of a facility or system are not eligible for hazard mitigation funding. Hazard mitigation measures apply only to the damaged elements of a facility or system. For example:

- If flooding inundates a sanitary sewer and blocks manholes with sediment, cost-effective mitigation to prevent the blockage of the damaged manholes in a future disaster may be considered.

- If five columns (of a building with twenty columns) were damaged in an earthquake, reinforcing those five damaged columns might be a cost-effective hazard mitigation measure.

- If wind blew out some of the windows in a building, providing shutters for those windows only (not all windows) could be a cost-effective mitigation measure.

- If a wing of a building was damaged, mitigation measures would apply only to that wing, not the whole building.

Undamaged facilities are not eligible for Section 406 hazard mitigation funding. However, they may be eligible for hazard mitigation funding under Section 404 (Stafford Act). Section 404 funding is managed by the State and does not fall under the jurisdiction of the Public Assistance Program. Measures under Section 404 do not have to be structural in nature. Questions regarding the Section 404 hazard mitigation program should be directed to your State Hazard Mitigation Officer.

What do I need to do to obtain Section 406 hazard mitigation funding?

- If your project is eligible for permanent repair, look for, and request your project be evaluated for, hazard mitigation opportunities.

- Submit benefit-cost documentation for each mitigation proposal.

APPENDIX D
Historic Preservation

When providing funding under the Stafford Act, FEMA is required to comply with applicable Federal historic preservation laws and regulations, including the National Historic Preservation Act (NHPA).

Section 106 of the NHPA requires that FEMA consider the direct or indirect effects of projects it funds (referred to as undertakings) on historic properties. Historic properties include pre-historic and historic districts, sites, buildings, structures, and objects listed on, or eligible for inclusion on, the National Register of Historic Places (NRHP). Historic properties also include those of traditional religious and cultural importance to an Indian tribe that meet the National Register of Historic Places criteria. Facilities can generally be considered historic when they are 50 years of age or older. However, historic properties are not limited to buildings or well-known historic sites. Facilities as diverse as bridges, roads, industrial plants, landscapes, and areas once inhabited by prehistoric populations may be considered historic properties.

Many Public Assistance projects have the potential to affect historic properties. These projects include:

- Repair, restoration, mitigation, or demolition of historic buildings, structures, or objects

- Repair, restoration, mitigation, and demolition projects in pre-historic and historic districts

- Improved, alternate, mitigation, or relocation projects affecting undisturbed areas that may contain pre-historic or historic sites

FEMA must ensure the following steps are accomplished before approving funding:

- Determination of the Area of Potential Effects

- Identification of historic properties

- Evaluation of the effects of the proposed projects on historic properties

- Resolution of adverse effects on historic properties
- Consultation with the State Historic Preservation Office (SHPO) or Tribal Historic Preservation Office (THPO) and other interested parties

In some cases, FEMA may establish a programmatic agreement with the Advisory Council on Historic Preservation (ACHP) and the SHPO or the THPO at the beginning of the disaster recovery process to address projects potentially falling within the scope of the NHPA and to expedite the Section 106 compliance process. In other cases, FEMA may have already established a programmatic agreement with the ACHP and SHPO or THPO that applies to current and future disasters. Your FEMA PAC Crew Leader and historic Technical Specialist will guide you through the historic preservation requirements.

What do I need to do?

- Help your FEMA PAC Crew Leader identify historic preservation issues related to your project. If you are preparing your own Subgrant Applications (*Project Worksheets*) for small projects, include historic preservation issues on your Subgrant Application (*Project Worksheet*).
- Be prepared submit the following with your Subgrant Applications (*Project Worksheets*):
  - Site plans, drawings, or sketches
  - Design/construction plans or drawings
  - Record of contact with regulatory agencies and other interested parties (e.g., the local historic society)
  - Copies of existing permits and permit applications
  - Information on the age of the facility, especially whether a facility or area is on the NRHP
  - Dates of construction and modification of the facility
  - Photographs of the damaged facility and surrounding area
  - Photographs of any historic markers or plaques
- Remember that failure to comply with historic preservation requirements will jeopardize FEMA funding for your project.

APPENDIX E
Environmental Compliance

When providing assistance under the Stafford Act, FEMA must comply with applicable Federal environmental laws and their implementing regulations and Executive Orders. The Federal environmental laws that most often relate to FEMA-funded projects include: the Clean Water Act, the Clean Air Act, the Coastal Barriers Resources Act, the Coastal Zone Management Act, the Resources Recovery and Conservation Act, the Endangered Species Act (ESA), and the National Environmental Policy Act (NEPA). The three Executive Orders most frequently encountered in FEMA projects address: floodplain management (Executive Order 11988), wetland protection (Executive Order 11990), and environmental justice (Executive Order 12898). While all projects must comply with these laws and Executive Orders, the location and nature of the project determines whether or not a law is relevant. Some of these laws have exclusions or allow for expedited consultations for certain types of work. FEMA will apply these mechanisms to clearance for projects where appropriate. FEMA and the States also have established procedures to expedite compliance of those projects that do not fall under the exceptions. Early identification of factors that affect compliance is the key to expedited review and approval.

NEPA requires that Federal agencies include environmental planning in their decisionmaking process, as well as mission objective, cost, and technological issues. FEMA complies with this requirement by evaluating the potential environmental impacts of a proposed project. NEPA also requires Federal agencies to ensure appropriate public involvement.

Over 75 percent of projects funded by the Public Assistance Program are either emergency activities or projects to repair facilities to their pre-disaster design. These types of actions are statutorily excluded from the NEPA review process. If the damaged facility is proposed to be upgraded or improved or if mitigation is being added, FEMA must conduct and document the NEPA review. Generally, FEMA must complete this NEPA review before the project is initiated, otherwise funding may be jeopardized. Regardless of the level of review required by NEPA, other relevant environmental laws and Executive Orders must be addressed.

Special consideration should be given to compliance with the ESA (Section 7). Unlike NEPA, emergency actions are not exempt from ESA. FEMA must consult with either the U.S. Fish and Wildlife Service (USFWS) or the National Marine Fisheries Service (NMFS) to determine the potential to affect protected species and their habitats, even for emergency actions. However, there are provisions in the ESA for emergency consultation. If you are taking any action which has the potential to affect threatened or endangered species or their habitats, you must notify FEMA and the appropriate regulatory agency (USFWS or NMFS) prior to taking the emergency action. This notification can be in the form of a phone call, an email, or a fax. This notification is required. If USFWS or NMFS is not notified of the emergency action and it potentially affects protected species or their habitats, FEMA cannot fund the project. After the emergency, FEMA can conclude the consultation process with USFWS or NMFS.

A "Yes" response to any of the *Special Consideration Questions* from the FEMA PAC Crew Leader is an indication that a more detailed review for compliance with environmental requirements might be triggered. If your project is near or affects a stream, a wetland, or other body of water, requires the temporary or permanent destruction of an area of natural vegetation, or is in or near a special resource area, like a wildlife refuge, forest, or park, an environmental compliance review will likely be required.

Your FEMA PAC Crew Leader will assign an environmental Technical Specialist to assist you.

What is my role?

- While it generally is the FEMA PAC Crew Leader's responsibility to complete the *Special Consideration Questions* form, he or she needs your help in identifying Special Considerations related to your project. The FEMA PAC Crew Leader will ask you a standard set of questions to help identify these issues.

- Be prepared to submit the following with your Subgrant Applications (*Project Worksheets*):
  - Site plans, drawings or sketches
  - Design/construction plans or drawings
  - Hydraulic/hydrological study or analysis
  - Location and site maps
  - Flood Insurance Rate Maps

- Record of contact with regulatory agencies
- Copies of existing permits and permit applications
- Photographs of the damaged facility and surrounding area

▶ If you are preparing your own Subgrant Applications (*Project Worksheets*), address Special Consideration issues on your Subgrant Applications (*Project Worksheets*), as appropriate.

▶ Submit projects with identified Special Consideration issues as soon as possible, as FEMA must complete the review for environmental compliance prior to initiation of the project work.

▶ Keep a copy of the *Special Consideration Questions* form with the other documentation on the project to show that regulatory issues were considered.

▶ Use a separate paragraph within the **SCOPE OF WORK** block on the Subgrant Application (*Project Worksheet*) to describe any proposed changes to the pre-disaster design of the facility and any proposed improved project, alternate project, relocation, hazard mitigation measure, or upgrade to applicable codes and standards.

▶ You are responsible for complying with all applicable environmental laws regardless of the funding source. Examples of other environmental laws include the Clean Water Act (Sections 401, 402, and 404), the Clean Air Act (New Source Review), state endangered species acts, and local noise ordinances. As a condition of funding, FEMA and the State will require that you obtain all applicable permits and document compliance with all applicable environmental laws.

▶ Sometimes local, State, or Federal regulations have permit exemptions. Contact the appropriate agencies about any exemptions and expedited permit processes that may be applicable. FEMA may withdraw funding for a project if you do not obtain the appropriate permits or comply with applicable environmental laws, regulations, and Executive Orders.

FORM

FEMA Form 90-120 *Special Consideration Questions*

(Forms are available at **www.fema.gov/government/grant/pa/forms.shtm.**)

APPENDIX F
FEMA Forms

Sample copies of forms follow this page and are available on the FEMA Web site (**www.fema.gov/government/grant/pa/forms.shtm**).

An authorized representative of the applicant must sign each page of the forms to certify the accuracy of the information provided.

FEMA Form 90-49	Request for Public Assistance (Pre-application)
FEMA Form 90-61	Hazard Mitigation Proposal (HMP)
FEMA Form 90-91	Project Worksheet (Subgrant Application)
FEMA Form 90-91A	Project Worksheet – Damage Description and Scope of Work Continuation Sheet
FEMA Form 90-91B	Project Worksheet – Cost Estimate Continuation Sheet
FEMA Form 90-91C	Project Worksheet – Maps and Sketches Sheet
FEMA Form 90-91D	Project Worksheet – Photo Sheet
FEMA Form 90-118	Validation Worksheet
FEMA Form 90-119	Project Validation Form
FEMA Form 90-120	Special Consideration Questions
FEMA Form 90-121	Private Nonprofit (PNP) Facility Questionnaire
FEMA Form 90-122	Historic Review Assessment for Determination of Adverse Effect
FEMA Form 90-123	Force Account Labor Summary Record
FEMA Form 90-124	Materials Summary Record
FEMA Form 90-125	Rented Equipment Summary Record
FEMA Form 90-126	Contract Work Summary Record
FEMA Form 90-127	Force Account Equipment Summary Record
FEMA Form 90-128	Applicant's Benefits Calculation Worksheet

DEPARTMENT OF HOMELAND SECURITY
FEDERAL EMERGENCY MANAGEMENT AGENCY
REQUEST FOR PUBLIC ASSISTANCE

O.M.B. No. 1660-0017
Expires October 31, 2008

PAPERWORK BURDEN DISCLOSURE NOTICE

Public reporting burden for this form is estimated to average 10 minutes. Burden means the time, effort and financial resources expended by persons to generate, maintain, disclose, or to provide information to us. You may send comments regarding the burden estimate or any aspect of the collection, including suggestions for reducing the burden to: Information Collections Management, U.S. Department of Homeland Security, Federal Emergency Management Agency, 500 C Street, SW, Washington, DC 20472, Paperwork Reduction Project (OMB Control Number 1660-0017). You are not required to respond to this collection of information unless a valid OMB number appears in the upper right corner of this form. **NOTE: Do not send your completed questionnaire to this address.**

APPLICANT *(Political subdivision or eligible applicant.)*

DATE SUBMITTED

COUNTY *(Location of Damages. If located in multiple counties, please indicate.)*

APPLICANT PHYSICAL LOCATION

STREET ADDRESS

CITY | COUNTY | STATE | ZIP CODE

MAILING ADDRESS *(If different from Physical Location)*

STREET ADDRESS

POST OFFICE BOX | CITY | STATE | ZIP CODE

Primary Contact/Applicant's Authorized Agent | **Alternate Contact**

NAME | NAME

TITLE | TITLE

BUSINESS PHONE | BUSINESS PHONE

FAX NUMBER | FAX NUMBER

HOME PHONE *(Optional)* | HOME PHONE *(Optional)*

CELL PHONE | CELL PHONE

E-MAIL ADDRESS | E-MAIL ADDRESS

PAGER & PIN NUMBER | PAGER & PIN NUMBER

Did you participate in the Federal/State Preliminary Damage Assessment (PDA)? ☐ Yes ☐ No

Private Non-Profit Organization? ☐ Yes ☐ No
If yes, which of the facilities identified below best describe your organization? _____

Title 44 CFR, part 206.221(e) defines an eligible private non-profit facility as:"... any private non-profit educational, utility, emergency, medical or custodial care facility, including a facility for the aged or disabled, and other facility providing essential governmental type services to the general public, and such facilities on Indian reservations." "Other essential governmental service facility means museums, zoos, community centers, libraries, homeless shelters, senior citizen centers, rehabilitation facilities, shelter workshops and facilities which provide health and safety services of a governmental nature. All such facilities must be open to the general public."

Private Non-Profit Organizations must attach copies of their Tax Exemption Certificate and Organization Charter or By-Laws. If your organization is a school or educational facility, please attach information on accreditation or certification.

Official Use Only: FEMA-_____-DR-_____-____ FIPS# _____ Date Received: _____

FEMA Form 90-49, FEB 06   REPLACES ALL PREVOUS EDITIONS.

DEPARTMENT OF HOMELAND SECURITY
FEDERAL EMERGENCY MANAGEMENT AGENCY
HAZARD MITIGATION PROPOSAL (HMP)

Sheet _____ of _____ Sheets

NAME OF APPLICANT	CATEGORY	DSR NUMBER

SCOPE OF MITIGATION WORK:

ESTIMATE OF WORK

QUANTITY	UNIT	MATERIAL AND/OR DESCRIPTION	UNIT	COST (Dollars)
			TOTAL (Not to be included in DSR)	

RECOMMENDED BY (Signature)*	AGENCY	DATE
CONCURRENCE BY STATE INSPECTOR (Signature)*	AGENCY	DATE
CONCURRENCE BY LOCAL REPRESENTATIVE (Signature)*	AGENCY	DATE

NOTE: *Signature by the Federal Inspector is not an approval of this work, and signature by the State and local applicant is not a commitment to perform the work.

FEMA Form 90-61, JUL 07   REPLACES ALL PREVIOUS EDITIONS.

SAMPLE

U.S. DEPARTMENT OF HOMELAND SECURITY
FEDERAL EMERGENCY MANAGEMENT AGENCY
PROJECT WORKSHEET

O.M.B. No. 1660-0017
Expires October 31, 2008

PAPERWORK BURDEN DISCLOSURE NOTICE

Public reporting burden for this form is estimated to average 90 minutes per response. Burden means the time, effort and financial resources expended by persons to generate, maintain, disclose, or to provide information to us. You may send comments regarding the burden estimate or any aspect of the collection, including suggestions for reducing the burden to: Information Collections Management, U.S. Department of Homeland Security, Federal Emergency Management Agency, 500 C Street, SW, Washington, DC 20472, Paperwork Reduction Project (OMB Control Number 1660-0017). You are not required to respond to this collection of information unless a valid OMB number appears in the upper right corner of this form. **NOTE: Do not send your completed questionnaire to this address.**

DISASTER FEMA- /DR-	PROJECT NO.	PA ID NO.	DATE	CATEGORY
DAMAGED FACILITY			WORK COMPLETE AS OF	%
APPLICANT		COUNTY		
LOCATION			LATITUDE	LONGITUDE

DAMAGE DESCRIPTION AND DIMENSIONS

SCOPE OF WORK

Does the Scope of Work change the pre-disaster conditions at the site? ☐ Yes ☐ No
Special Considerations issues included? ☐ Yes ☐ No   Hazard Mitigation proposal included? ☐ Yes ☐ No
Is there insurance coverage on this facility? ☐ Yes ☐ No

PROJECT COST

ITEM	CODE	NARRATIVE	QUANTITY/UNIT	UNIT PRICE	COST
			TOTAL COST ▶		

PREPARED BY	TITLE	SIGNATURE
APPLICANT REP.	TITLE	SIGNATURE

FEMA Form 90-91, FEB 06   REPLACES ALL PREVIOUS EDITIONS.

F-4   Appendix F: FEMA Forms

U.S. DEPARTMENT OF HOMELAND SECURITY FEDERAL EMERGENCY MANAGEMENT AGENCY **PROJECT WORKSHEET** - Damage Description and Scope of Work Continuation Sheet				O.M.B. No. 1660-0017 Expires October 31, 2008
DISASTER FEMA-_____DR-____	PROJECT NO.	PA ID NO.	DATE	CATEGORY
APPLICANT		COUNTY		

SAMPLE

PREPARED BY:	TITLE:

FEMA Form 90-91A, FEB 06

Appendix F: FEMA Forms

U.S. DEPARTMENT OF HOMELAND SECURITY
FEDERAL EMERGENCY MANAGEMENT AGENCY
PROJECT WORKSHEET - Cost Estimate Continuation Sheet

O.M.B. No. 1660-0017
Expires October 31, 2008

DISASTER	PROJECT NO.	PA ID NO.	DATE	CATEGORY
APPLICANT		COUNTY		

PROJECT COST

ITEM	CODE	NARRATIVE	QUANTITY/UNIT	UNIT PRICE	COST

TOTAL COST ▶

PREPARED BY:	TITLE:

FEMA Form 90-91B, FEB 06

Appendix F: FEMA Forms

U.S. DEPARTMENT OF HOMELAND SECURITY **FEDERAL EMERGENCY MANAGEMENT AGENCY** **PROJECT WORKSHEET - Maps and Sketches Sheet**				O.M.B. No. 1660-0017 Expires October 31, 2008
DISASTER FEMA _____ DR ____	PROJECT NO.	PA ID NO.	DATE	CATEGORY
APPLICANT		COUNTY		

SAMPLE

FEMA Form 90-91C, FEB 06

F-8 — Appendix F: FEMA Forms

DEPARTMENT OF HOMELAND SECURITY
FEDERAL EMERGENCY MANAGEMENT AGENCY
VALIDATION WORKSHEET

DISASTER:
FEMA-   -DR-

APPLICANT	PA ID NO	PROJECT WORKSHEET NO
SPECIALIST	AGENCY	TELEPHONE NO

I- GENERAL - ALL PROJECTS

VALIDATION ITEM	REMARKS
☐ Review projects ☐ Visit site ☐ Statement of work 　　☐ Accurate 　　☐ Complete 　　☐ Eligible ☐ Pictures ☐ Sketches/drawings	

II- COMPLETED WORK

☐ Forced Account Labor 　　☐ Eligible employee 　　☐ Hours 　　　　☐ Regular 　　　　☐ Overtime ☐ Fringe benefits 　　☐ Regular 　　☐ Overtime ☐ Calculations	

III- FORCE ACCOUNT EQUIPMENT

☐ Labor hours exceeds or match Equipment hours ☐ FEMA rates used ☐ PAC approved rates used ☐ Mileage used for automobiles, busses, pickups, and ambulances ☐ Calculations	

IV- LEASED/RENTAL EQUIPMENT

☐ Invoice ☐ Price reasonable ☐ Operation/labor cost ☐ Gasoline/oil/lubricants ☐ Eligible repairs/parts ☐ Calculations	

V- MATERIALS

☐ Purchase orders/invoices ☐ Inventory records/stock tickets ☐ Calculations	

FEMA Form 90-118, NOV 98

VI- CONTRACT	
VALIDATION ITEM	REMARKS
☐ Price reasonable ☐ Competitive bids ☐ Exception ☐ Follow procurement procedures ☐ Calculations	

VII- WORK TO BE COMPLETED	
☐ Cost estimating method approved by PAC ☐ Calculations	

VIII- SPECIAL CONSIDERATIONS	
☐ Insurance ☐ Mitigation ☐ Environmental ☐ Historic	

ADDITIONAL REMARKS

FEMA Form 90-118, NOV 98, Back

DEPARTMENT OF HOMELAND SECURITY
FEDERAL EMERGENCY MANAGEMENT AGENCY
PROJECT VALIDATION FORM

DISASTER: FEMA- -DR-

APPLICANT	DATE	PA ID NO.
SPECIALIST	AGENCY	
CONTACT PERSON	TELEPHONE NO.	

The projects listed below were validated from:
☐ Sample 1 C.V.   ☐ Sample 1 and 2 C.V.

VALIDATION

A Project Worksheet No.	B Applicant Estimate	C Eligibility Variance	D Cost Estimate Variance	E Comments
	$	$	$	
SUBTOTAL	B $	C $	D $	PERCENT OF VARIANCE %
TOTAL VARIANCE (COL. C + D) = F			F $	(F divided by B)

II-VALADIATION RESULTS

☐ VARIANCE WITHIN 20% 1st VALIDATION   ☐ VARIANCE WITHIN 20% 2nd VALIDATION
☐ VARIANCE WITHIN 20% 1st & 2nd VALIDATION

III-RECOMMENDATION

☐ APPROVE FUNDING, VARIANCE WITHIN 20%   ☐ PROVIDE TECHNICAL ASSISTANCE, VARIANCE EXCEEDS 20%

FEMA Form 90-119, NOV 98

DEPARTMENT OF HOMELAND SECURITY
FEDERAL EMERGENCY MANAGEMENT AGENCY
SPECIAL CONSIDERATION QUESTIONS

O.M.B. NO. 1660-0017
Expires October 31, 2008

APPLICANT	PA ID NO.	DATE
PROJECT NAME	PROJECT NO.	LOCATION

Form must be filledout - for each project.

1. Does the damaged facility or item of work have insurance and/or is it an insurable risk? (e.g., buildings, equipment, vehicles, etc.)
   ☐ Yes ☐ No ☐ Unsure
   Comments

2. Is the damaged facility located within a floodplain or coastal high hazard area/or does it have an impact on a floodplain or wetland?
   ☐ Yes ☐ No ☐ Unsure
   Comments

3. Is the damaged facility or item of work located within or adjacent to a Coastal Barrier Resource System Unit or an Otherwise Protected area?
   ☐ Yes ☐ No ☐ Unsure
   Comments

4. Will the proposed facility repairs/reconstruction change the pre-disaster condition? (e.g., footprint, material, location, capacity, use or function)
   ☐ Yes ☐ No ☐ Unsure
   Comments

5. Does the applicant have a hazard mitigation proposal or would the applicant like technical assistance for a hazard mitigation proposal?
   ☐ Yes ☐ No ☐ Unsure
   Comments

6. Is the damaged facility on the National Register of Historic Places or the state historic listing? Is it older than 50 years? Are there other, similar buildings near the site? ☐ Yes ☐ No ☐ Unsure
   Comments

7. Are there any pristine or undisturbed areas on, or near, the project site? Are there large tracts of forestland?
   ☐ Yes ☐ No ☐ Unsure
   Comments

8. Are there any hazardous materials at or adjacent to the damaged facility and/or item of work?
   ☐ Yes ☐ No ☐ Unsure
   Comments

9. Are there any other environmental or controversial issues associated with the damaged facility and/or item of work?
   ☐ Yes ☐ No ☐ Unsure
   Comments

FEMA Form 90-120, FEB 96   PREVIOUS EDITION OBSOLETE

DEPARTMENT OF HOMELAND SECURITY
FEDERAL EMERGENCY MANAGEMENT AGENCY
PNP FACILITY QUESTIONNAIRE

O.M.B. NO. 1660-0017
Expires October 31, 2008

PAPERWORK BURDEN DISCLOSURE NOTICE

Public reporting burden for this form is estimated to average 30 minutes per response. The burden estimate includes the time for reviewing instructions, searching existing data sources, gathering and maintaining the needed data, and completing, reviewing, and submitting the form. You are not required to respond to this collection of information unless a valid OMB control number appears in the upper right corner of this form. Send comments regarding the accuracy of the burden estimate and any suggestions for reducing this burden to: Information Collections Management, Department of Homeland Security, Federal Emergency Management Agency, 500 C Street, SW, Washington, DC, 20472, Paperwork Reduction Project (1660-0017). **Please do not send your completed survey to the above address.**

FEMA and State personnel will use this questionnaire to determine the eligibility of specific facilities of an approved Private Non-Profit (PNP) organization (See 44 CFR 206.221). Owners of critical facilities (i.e., power, water (including providing by an irrigation organization or facility; if it is not provided solely for irrigation purposes), sewer, wastewater treatment, communications and emergency medical care) can apply directly to FEMA for assistance for emergency work (debris removal and emergency protective measures) and permanent work (repair, restore or replace a damaged facility). Owners of non-critical facilities can apply directly to FEMA for assistance for emergency work, but must first apply to the U. S. Small Business Administration (SBA) for assistance for permanent work. If the owner of a non-critical facility does not qualify for an SBA loan or the cost to repair the damaged facility exceeds the SBA loan amount, the owner may apply to FEMA for assistance.

1. Name of PNP Organization

2. Name of the damaged facility and location

3. What was the primary purpose of the damaged facility?

4. Is the facility a critical facility as described above? ☐ Yes ☐ No

5. Who may use the facility

6. What fee, if any, is charged for the use of the facility

7. Was the facility in use at the time of the disaster? ☐ Yes ☐ No

8. Did the facility sustain damage as a direct result of the disaster? ☐ Yes ☐ No

9. What type of assistance is being requested?

10. Does the PNP organization own the facility? ☐ Yes ☐ No

11. If "Yes", obtain proof of ownership; check here if attached ☐

12. Does the PNP organization have the legal responsibility to repair the facility? ☐ Yes ☐ No

13. If "Yes", provide proof of legal responsibility, check here if attached. ☐ Yes ☐ No

14. Is the facility insured? ☐ Yes ☐ No

15. If "Yes", obtain a copy of the insurance policy, check here if attached ☐

Additional information or comments:

CONTACT PERSON   DATE

FEMA Form 90-121, FEB 06

DEPARTMENT OF HOMELAND SECURITY
FEDERAL EMERGENCY MANAGEMENT AGENCY
HISTORIC REVIEW ASSESSMENT FOR DETERMINATION OF ADVERSE EFFECT

PA ID NO.	PROJECT NO.	LATITUDE/LONGITUDE
ADDRESS/LOCATION OF FACILITY/SITE		HISTORIC NAME AND ID #

HISTORIC STATUS ☐ NHL ☐ NR/NR eligible ☐ State Register or other ☐ Contributing to Historic District

1. Describe disaster damage, particularly as it relates to character-defining features:

2. The proposed scope of work will (check all that apply):
   ☐ Repair or replace non character-defining features
   ☐ Alter or remove historic features/elements
   ☐ Disturb, destroy or make archeological resources
   ☐ Other
   ☐ Repair and or replace historic features/elements, in kind to return facility to pre-disaster condition
   ☐ Add non-historic features/elements to a historic facility, setting or
   ☐ Include mitigation, an alternate project or an improved project

3. Describe measures to prevent or minimize loss or impairment of character-defining features:

4. Attachments:
   ☐ Maps   ☐ Field Notes   ☐ Scope of Work   ☐ Site Plan   ☐ National Register Nomination Form
   ☐ Drawings   ☐ Research Material   ☐ Project Worksheet   ☐ Specifications   ☐ Summary Views of Interested Parties
   ☐ Photographs ☐ Archeological   ☐ Other

5. Conclusions:
   ☐ 5a. No Character-defining features will be affected.
   ☐ 5b. The above action(s) meets the conditions for a Programmatic Exclusion # _____ of the Programmatic Agreement governing historic review.
   ☐ 5c. The above action(s) substantially conforms with the applicable parts of the Secretary of Interior's Standards and Guidelines for Archeology and Historic Preservation.
   ☐ 5d. Further consultation with the SHPO and applicant in accordance with the Programmatic Agreement is required.
   ☐ 5e. Development of STMA or Memorandum of Agreement is required to treat the adverse effect.

6. Assessment of Adverse Effect (check one):   ☐ No Adverse Effect   ☐ Adverse Effect

7. Specialist: Your signature shows that you have reviewed this form and related material for conformity with requirements in FEMA's Programmatic Agreement governing compliance with the National Historic Preservation Act, applicable parts of the Secretary of the Interior's Standards for Rehabilitation and Guidelines for Rehabilitating Historic Buildings 1992 (Standards), the Secretary of the Interior's Guidelines for Archeological Documentation (Guidelines), or any other applicable Secretary of the Interior's Standards, OR 44 CFR Part 206, and FEMA Management Policies, and have provided your best professional opinion.

COMMENTS

NAME	FIELD OF EXPERTISE	DATE

8. Action Taken and Date

FEMA Form 90-122, NOV 98

Appendix F: FEMA Forms

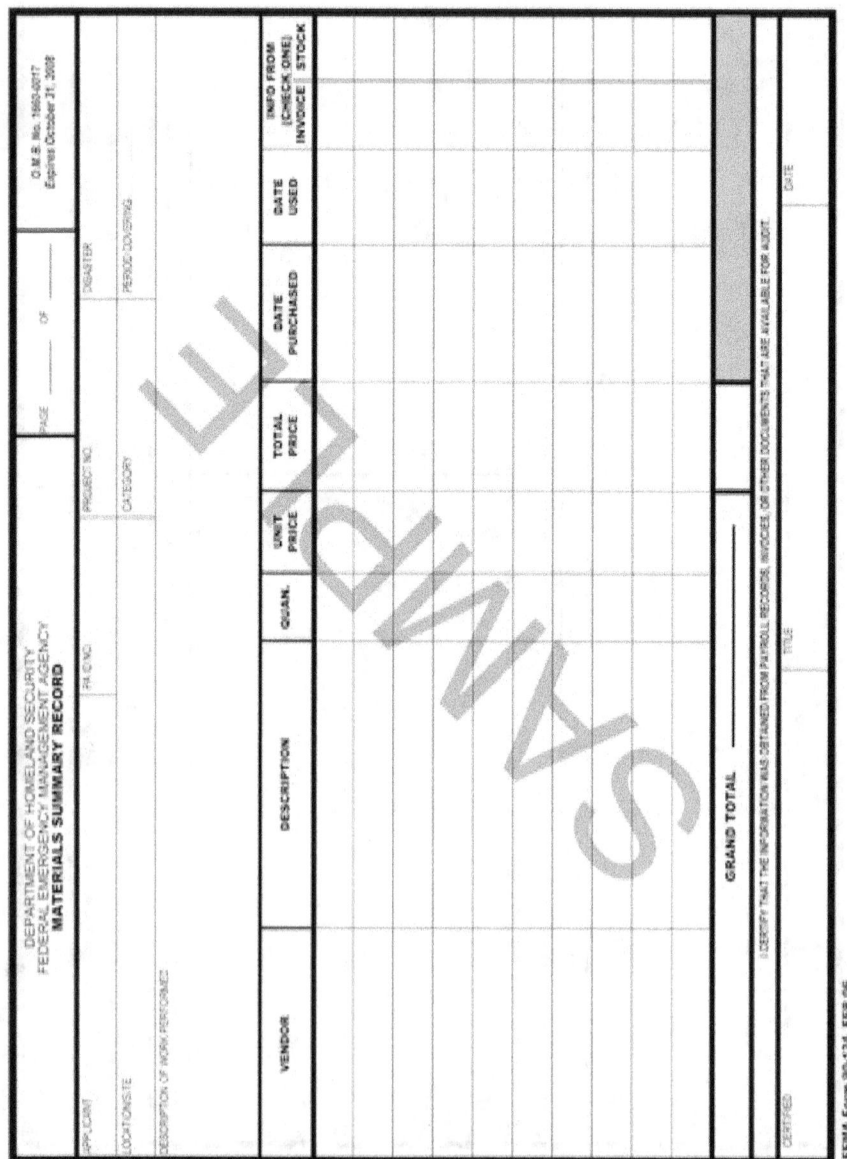

Appendix F: FEMA Forms **F-17**

Appendix F: FEMA Forms

DEPARTMENT OF HOMELAND SECURITY
FEDERAL EMERGENCY MANAGEMENT AGENCY
FORCE ACCOUNT EQUIPMENT SUMMARY RECORD

O.M.B. No. 1660-0017
Expires October 31, 2008

APPLICANT | PA ID NO. | PROJECT NO. | PAGE ___ OF ___ | DISASTER

LOCATION/SITE | CATEGORY | PERIOD COVERING

DESCRIPTION OF WORK PERFORMED

TYPE OF EQUIPMENT	EQUIPMENT CODE NUMBER	OPERATOR'S NAME	DATES AND HOURS USED EACH DAY							COSTS		
INDICATE SIZE, CAPACITY, HOURSEPOWER, MAKE AND MODEL AS APPROPRIATE			DATE							TOTAL HOURS	EQUIPMENT RATE	TOTAL COST
			HOURS									
			HOURS									
			HOURS									
			HOURS									
			HOURS									
			HOURS									

GRAND TOTAL

I CERTIFY THAT THE ABOVE INFORMATION WAS OBTAINED FROM PAYROLL RECORDS, INVOICES, OR OTHER DOCUMENTS THAT ARE AVAILABLE FOR AUDIT.

CERTIFIED | TITLE | DATE

FEMA Form 90-127, FEB 06

DEPARTMENT OF HOMELAND SECURITY
FEDERAL EMERGENCY MANAGEMENT AGENCY
APPLICANT'S BENEFITS CALCULATION WORKSHEET

PAGE _____ OF _____

O.M.B. No. 1660-0017
Expires October 31, 2008

APPLICANT:

PA ID NO:

DISASTER:

PROJECT NO:

FRINGE BENEFITS (by %)	REGULAR TIME	OVERTIME
HOLIDAYS		
VACATION LEAVE		
SICK LEAVE		
SOCIAL SECURITY		
MEDICARE		
UNEMPLOYMENT		
WORKER'S COMP.		
RETIREMENT		
HEALTH BENEFITS		
LIFE INS. BENEFITS		
OTHER		
TOTAL in % annual salary		

COMMENTS:

I CERTIFY THAT THE INFORMATION ABOVE WAS TRANSCRIBED FROM PAYROLL RECORDS OR OTHER DOCUMENTS WHICH ARE AVAILABLE

Name	TITLE	DATE

FEMA Form 90-128, FEB 06

APPENDIX G
Glossary of Terms

Applicant

A State agency, local government, Indian Tribe, authorized tribal organization, Alaska Native village or organization and certain Private Nonprofit (PNP) organizations that submit a request for disaster assistance under the Presidentially declared major disaster or emergency. The terms "applicant" and subgrantee" are often used interchangeably.

Applicants' Briefing

A meeting conducted by a representative of the State for potential Public Assistance applicants. The briefing occurs after an emergency or major disaster has been declared and addresses Public Assistance application procedures, administrative requirements, funding, and program eligibility criteria.

Applicant Liaison

A State representative responsible for providing applicants with State-specific information and documentation requirements. The Applicant Liaison works closely with the PAC Crew Leader to provide any technical assistance or guidance the applicant may require. The terms "Applicant Liaison" and "State Public Assistance (PA) Representative" are often used interchangeably.

Case Management

A system approach to provision of equitable and timely service to applicants for disaster assistance. Organized around the needs of the applicant, the system consists of a single point of coordination, a team of on-site specialists, and a centralized, automated filing system.

Case Management File (CMF)

A centralized data bank of all applicant activities. Data entered into this bank creates a chronological history of everything that has taken place with an applicant from the time they apply for assistance until they have received all monies and their file has been closed.

Closeout

Grant closure occurs when FEMA determines that all applicable administrative actions related to the Public Assistance Program are complete and all programs funds are reconciled. At this stage, all Public Assistance Program projects have been completed, the State has awarded all grant funds and submitted its final expenditure report to FEMA, and FEMA has adjusted the funding level for the program as appropriate.

Contractor

Any individual, partnership, corporation, agency, or other entity (other than an organization engaged in the business of insurance) performing work by contract for the Federal Government or a State or local agency, or Tribal government.

Cost Estimating Format (CEF)

A forward-pricing methodology for estimating the total cost of repair for large permanent projects by use of construction industry standards. The format uses a base cost estimate and design and construction contingency factors, applied as a percentage of the base cost.

Declarations

There are two types of declarations (Emergency Declarations and Major Disaster Declarations). Both declaration types authorize the President to provide Federal disaster assistance. However, the cause of the declaration and the type and amount of assistance differ.

Designated Area

Any emergency- or major disaster-affected portion of a State that has been determined eligible to apply for Federal assistance.

Emergency

Any occasion or instance for which, in the determination of the President, Federal assistance is needed to supplement State and local efforts and capabilities to save lives and to protect property and public health and safety, or to lessen or avert the threat of a catastrophe in any part of the United States.

Emergency Work

That work which is performed to reduce or eliminate an immediate threat to life, to protect health and safety, and to protect improved property that is threatened in a significant way as a result of a major disaster. Emergency work frequently includes clearance and removal of debris and temporary restoration of essential public facilities and services (Categories A and B).

Executive Orders (EOs)

Legally binding orders given by the President to Federal Administrative Agencies. Executive Orders are generally used to direct Federal agencies and officials in their execution of congressionally established laws or policies. Executive Orders do not require congressional approval to take effect, but they have the same legal weight as laws passed by Congress.

Expedited Payments

An advance of grants to assist with payment of emergency work after a disaster event. The amount of funding is 50 percent of the Federal share of emergency costs as identified during the Preliminary Damage Assessment. Payment for Category A will be made within 60 days after the estimate was made and no later than 90 days after the Pre-application (*Request for Public Assistance*) was submitted.

Facility

Any publicly or Private Nonprofit (PNP)-owned building, works, system, or equipment (built or manufactured) or certain improved and maintained natural features. Land used for agricultural purposes is not a facility.

Flood Control Works

Facilities constructed for the purpose of eliminating or reducing the threat of flood, e.g., levees, floodwalls, flood control channels and dams designed for flood control.

Force Account

Labor performed by the applicant's employees and applicant-owned equipment, rather than by a contractor.

FEMA-State Agreement

A formal legal document stating the understandings, commitments, and binding conditions for assistance applicable as the result of a major disaster or emergency declared by the President.

Grantee

The State, in most cases, acts as the grantee for the Public Assistance Program. The grantee is accountable for the use of the funds provided. The terms "grantee" and "State" are often used interchangeably.

Hazard Mitigation

Any cost-effective action taken to prevent or reduce the threat of future damage to a facility from a disaster event.

Immediate Needs Funding

An advance of grant funds for emergency work that must be performed immediately and paid for within the first 60 days after the major disaster declaration. The amount of funding is normally up to 50 percent of the Federal share of emergency costs.

Improved Property

A structure, facility, or item of equipment that was built, constructed, or manufactured. It includes improved and maintained natural features. Land used for agricultural purposes is not improved property.

Incident Period

The time interval during which the disaster-causing incident occurs. No Federal assistance under the Stafford Act shall be approved unless the damage or hardship to be alleviated resulted from the disaster-causing incident which took place during the incident period or was in anticipation of that incident.

Kickoff Meeting

The initial meeting of an applicant, the State PA Representative (Applicant Liaison), and the FEMA PAC Crew Leader. At this working session, the applicant provides a list of damages and receives comprehensive information about the Public Assistance Program and detailed guidance for the applicant's specific circumstances. This is the first step in establishing a partnership among FEMA, the State, and the applicant and is designed to focus on the specific needs of the applicant. The meeting focuses on the eligibility and documentation requirements that are most pertinent to the applicant.

Large Project

An eligible project, either emergency or permanent work, that has a damage dollar value at or above the fiscal year threshold. The threshold is adjusted each fiscal year to account for inflation. Large project funding is based on documented actual costs.

Major Disaster

Any natural catastrophe (including any hurricane, tornado, storm, high water, wind driven water, tidal wave, tsunami, earthquake, volcanic eruption, landslide, mudslide, snowstorm, or drought), or, regardless of cause, any fire, flood, or explosion, in any part of the United States, which in the determination of the President causes damage of sufficient severity and magnitude to warrant major disaster assistance under the Stafford Act to supplement the efforts and available resources of States, local governments, and disaster relief organizations in alleviating the damage, loss, hardship, or suffering caused thereby.

Mutual Aid Agreement

An agreement between jurisdictions or agencies to provide services across boundaries in an emergency or major disaster. Such agreements usually provide for reciprocal services or direct payment for services.

Obligated Funds

The funds FEMA makes available to the grantee for approved Public Assistance projects. The grantee is then required to make payment of the Federal share to the applicant as soon as practicable.

Other Essential Governmental Service Facilities

Private Nonprofit (PNP) museums, zoos, performing arts facilities, community arts centers, community centers, libraries, homeless shelters, senior citizen centers, rehabilitation facilities, mass transit facilities, shelter workshops and facilities which provide health and safety services of a governmental nature. All such facilities must be open to the general public.

PAC Crew Leader

A FEMA representative who works with the applicant to resolve disaster-related needs and to ensure that the applicant's projects are processed as efficiently and expeditiously as possible. The PAC Crew Leader ensures continuity of service throughout the delivery of the Public Assistance Program.

Permanent Work

That work which is required to restore a facility, through repairs or replacement, to its pre-disaster design, function, and capacity in accordance with applicable codes and standards (Categories C through G).

Pre-application

An applicant's official notification to FEMA of intent to apply for Public Assistance. The form provides general identifying information about the applicant. The terms "Pre-application" and "Request for Public Assistance" are often used interchangeably.

Preliminary Damage Assessment (PDA)

A survey performed to document the impact and magnitude of the disaster on individuals, families, businesses, and public property and to gather information for disaster management purposes. The information gathered is used to determine whether Federal assistance should be requested by the Governor and forms the basis for the disaster declaration request.

Private Nonprofit (PNP) Facilities

Educational, utility, irrigation, emergency, medical, rehabilitational, and temporary or permanent custodial care facilities and facilities on Indian reservations, as defined by the President. Other PNP facilities that provide essential services of a governmental nature are eligible and are listed in this Glossary of Terms under Other Essential Governmental Service Facilities.

Private Nonprofit (PNP) Organization

Any non-governmental agency or entity that currently has either an effective ruling letter from the U.S. Internal Revenue Service granting tax exemption or satisfactory evidence from the State that the non-revenue producing organization or entity is a nonprofit one organized or operating under State law.

Project Formulation

The process of identifying the eligible scope of work and estimating the costs associated with that scope of work for each applicant's projects.

Project Specialist

FEMA's specialist who works directly with the applicant in assessing damage sites and in developing scopes of work and cost estimates. The FEMA Project Specialist will also identify the need for other specialists and work with the FEMA PAC Crew Leader in obtaining their services for projects.

Project Worksheet

Form used to document the location, damage description and dimensions, scope of work, and cost estimate for a project. It is the basis for the grant. The terms "*Project Worksheet*" and "Subgrant Application" are often used interchangeably.

Public Assistance (PA)

Supplementary Federal assistance provided under the Stafford Act to State, local and Tribal governments or eligible PNPs to help them recover from Federally-declared major disasters and emergencies as quickly as possible.

Request for Public Assistance (RPA)

An applicant's official notification to FEMA of intent to apply for Public Assistance. The form provides general identifying information about the applicant. The terms "*Request for Public Assistance*" and "Pre-application" are often used interchangeably.

Small Project

An eligible project, either emergency or permanent work, that has a damage dollar value below the fiscal year threshold. The threshold is adjusted each fiscal year to account for inflation. Small project funding is based on estimated costs if actual costs are not yet available.

Special Considerations

Factors that must be addressed before Federal PA grant money can be obligated to repair or restore damaged facilities. These factors include, but are not limited to, general and flood insurance, historic preservation, environmental protection, and hazard mitigation.

Stafford Act

Robert T. Stafford Major Disaster Relief and Emergency Assistance Act (Stafford Act), PL 100-707, signed into law November 23, 1988; the 1988 law amended the Major Disaster Relief Act of 1974, PL 93-288. This Act constitutes the statutory authority for most Federal major disaster response activities especially as they pertain to FEMA and FEMA programs.

State Administrative Plan

The State is required to develop a State Administrative Plan to administer the Public Assistance Program. The Plan should include the designation of responsibilities for State agencies and include staffing for the Public Assistance Program. An approved State Administrative Plan must be on file with FEMA before grants will be approved for any major disaster. The approved State Administrative Plan should be incorporated into the State's emergency plan.

State Public Assistance (PA) Representative

An applicant's point of contact, designated by the State, who will help the applicant obtain FEMA assistance. The terms "State PA Representative" and "Applicant Liaison" are often used interchangeably.

Subgrantee

A State agency, local government, Indian Tribe, authorized tribal organization, Alaska Native village or organization, and certain Private Nonprofit organizations that submit a request for disaster assistance under the Presidentially declared major disaster or emergency. The terms "subgrantee" and "applicant" are often used interchangeably.

Subgrant Application

Form used to document the location, damage description and dimensions, scope of work, and cost estimate for a project. It is the basis for the grant. The terms "Subgrant Application" and "*Project Worksheet*" are often used interchangeably.

Technical Specialist

FEMA's Technical Specialist is a resource for the applicant. A Technical Specialist has a defined area of expertise, such as debris removal and disposal, roads and bridges, infrastructure, environmental and historic preservation compliance, insurance, cost estimating, or floodplain management.

APPENDIX H
FEMA Policies and Publications

9510 Public Assistance Program Administration and Appeals

9510.1   Coordination Requirements for Public Assistance and Fire Management Assistance Program Documentation and Appendix

9520 Public Assistance Eligibility

9521.1   Community Center Eligibility

9521.2   Private Nonprofit Museum Eligibility

9521.3   Private Nonprofit (PNP) Facility Eligibility

9521.4   Administering American Indian and Alaska Native Tribal Government Funding

9521.5   Eligibility of Charter Schools

9523.1   Snow Assistance Policy

9523.2   Eligibility of Building Inspections in a Post-Disaster Environment

9523.3   Provision of Temporary Relocation Facilities

9523.4   Demolition of Private Structures

9523.6   Mutual Aid Agreements for Public Assistance and Fire Management Assistance

9523.7   Public Assistance for Public Housing Facilities

9523.8   Mission Assignments for ESF #10

9523.9   100% Funding for Direct Federal Assistance and Grant Assistance

9523.10   Eligibility of Vector Control (Mosquito Abatement)

9523.11   Hazardous Stump Extraction and Removal Eligibility

9523.12   Debris Operations – Hand-Loaded Trucks and Trailers

9523.13   Debris Removal from Private Property

9523.15   Eligible Costs Related to Evacuations and Sheltering

9523.17	Emergency Assistance for Human Influenza Pandemic
9523.18	Host-State Evacuation and Sheltering Reimbursement
9523.19	Eligible Costs Related to Pet Evacuations and Sheltering
9523.20	Purchase and Distribution of Ice
9524.1	Welded Steel Moment Frame
9524.2	Landslides and Slope Failures
9524.3	Rehabilitation Assistance for Levees and other Flood Control Works Memo updating Policy 9524.3 – August 5, 2009
9524.4	Repair vs. Replacement of a Facility under 44 CFR §206.226(f) (The 50% Rule)
9524.5	Trees, Shrubs and Other Plantings Associated with Facilities
9524.6	Collections and Individual Object Eligibility
9524.7	Interim Welded Steel Moment Frame Policy for the Nisqually Earthquake Disaster
9524.8	Eligibility for Permanent Repair and Replacement of Roads on Tribal Lands
9524.9	Replacement of Animals Associated with Eligible Facilities
9524.10	Replacement of Equipment, Vehicles, and Supplies
9525.1	Post Disaster Property Tax Assessment
9525.2	Donated Resources
9525.3	Duplication of Benefits – Non-Government Funds
9525.4	Emergency Medical Care and Medical Evacuations
9525.5	Americans with Disabilities Act (ADA) Access Requirements
9525.6	Project Supervision and Management Costs of Subgrantees
9525.7	Labor Costs – Emergency Work
9525.8	Damage to Applicant Owned Equipment
9525.9	Section 324 Management Costs and Direct Administrative Costs
9525.11	Payment of Contractors for Grant Management Tasks
9525.12	Disposition of Equipment, Supplies and Salvageable Materials
9525.13	Alternate Projects
9525.14	Public Assistance Grantee Administrative Costs
9525.15	Telecommunications Support Lines for States
9525.16	Research-related Equipment and Furnishings

9526.1 Hazard Mitigation Funding Under Section 406 (Stafford Act)

9527.1 Seismic Safety – New Construction
9527.4 Construction Codes and Standards

9530 Public Assistance Insurance Requirements
9530.1 Retroactive Application of a Letter of Map Amendment (LOMA) or Letter of Map Revision (LOMR) to Public Assistance Grants

9550 Fire Suppression Assistance
9550.3 Interim Policy on Fire Suppression Assistance

9560 Compliance with Environmental, Historic, Cultural Initiatives, Other Laws
9560.1 Environmental Policy Memoranda
9560.3 Programmatic Agreement – Historic Review

9570 Standard Operating Procedures
9570.2 Standard Operating Procedure – Public Assistance Coordinator
9570.4 Standard Operating Procedure – Kickoff Meeting
9570.5 Standard Operating Procedure – Project Formulation
9570.6 Standard Operating Procedure – Validation of Small Projects
9570.7 Standard Operating Procedure – Immediate Needs Funding
9570.8 Standard Operating Procedure – Cost Estimating Format for Large Projects
9570.9 Standard Operating Procedure – Historic Review

9580 Job Aids and Fact Sheets
9580.2 Fact Sheet: Insurance Responsibilities for Field Personnel
9580.3 Fact Sheet: Insurance Considerations for Applicants
9580.4 Fact Sheet: Debris Operations: Clarification
9580.5 Fact Sheet: Elements of a Project Worksheet
9580.6 Fact Sheet: Electric Utility Repair (Public and Private Nonprofit)
9580.8 Fact Sheet: Eligible Sand Replacement on Public Beaches
9580.100 Fact Sheet: Mold Remediation

9580.101	Fact Sheet: 2006 Special Community Disaster Loan Program
9580.102	Fact Sheet: Relocation and Permanent Relocation Fact Sheet Clarification Memorandum
9580.103	Fact Sheet: GSA Disaster Recovery Purchasing Program
9580.104	Fact Sheet: Public Assistance for Ambulance Services
9580.201	Fact Sheet: Debris Removal – Applicant's Contracting Checklist
9580.202	Fact Sheet: Debris Removal – Authorities of Federal Agencies
9580.203	Fact Sheet: Debris Monitoring
9580.204	Fact Sheet: Documenting and Validating Hazardous Trees, Limbs, and Stumps

Other FEMA Publications

FEMA 321	Public Assistance Policy Digest
FEMA 322	Public Assistance Guide
FEMA 323	Public Assistance Applicant Handbook
FEMA 325	Public Assistance Debris Management Guide
FEMA 328	Public Assistance Program Brochure
FEMA 329S	El Programa Asistencia Publica

Other Sources

Robert T. Stafford Disaster Relief and Emergency Assistance Act, as amended

Title 44 of the Code of Federal Regulations

These publications are available on the FEMA Web site (**www.fema.gov/government/grant/pa/**).

Index

A

Administrative costs 18, H-2
Alternate projects 23, 26–28, 54, C-2, D-1, E-3, H-2
Appeals 25, 30, 50–51, 63
Applicants' Briefing 3, 7–9, 18, 21, 32, 62, G-1
Applicant Liaison 3, G-1, G-7
Audits 18, 32–33, 53, 58–59, 63

B

Bridges .... *(See Roads and Bridges)*
Buildings and Equipment (Category E) 15–16, 38, 49

C

Case Management File (CMF) 9, G-1
Categories of work 11, 13–17, 36, 38
Category A 13–14, 16, 38, 49, A-3–A-4, G-3
Category B 14–15, 38, 49, A-3–A-4
Category C 15, 38, 49
Category D 15, 38, 49
Category E 15–16, 38, 49
Category F 16, 38, 49
Category G 16–17, 38, 49
Category H 17
Closeout 3, 7–8, 33, 57–59, G-2
Community Development Districts A-5
Completion deadlines .... *(see Deadlines)*
Contracts 1, 10, 36, 43–45, 54
Cost estimates 4, 18, 22, 35, 41–47, G-2, G-6
Cost Estimating Format (CEF) 46–47, G-2, H-3
Cost overrun 30, 50, 53, 58
Cost share 1, 2, 6, 27
Critical PNP services A-3–A-4
Custodial Care Facilities (PNP) A-1–A-6

D

Damage description and dimensions 35, 39–40, G-6, G-8
Damaged facility 1, 23–28, 38–41, C-1, E-1
Deadlines 10–11, 24, 25, 30, 47-48, 49–50, 61–63
Debris Removal (Category A) 7–8, 10–11, 13–14, 38, 49, G-2, H-1, H-3, H-4
Declaration 3, 6, 8, 17–18, 37, 58, 61–62, G-2
Department of Commerce 2
Department of Housing and Urban Development (HUD) .... vii, 2
Department of Interior 2
Department of Transportation 2
Direct Federal Assistance 2, 6, H-1
Documentation 8, 11, 18–19, 21–22, 30, 32–33, 39, 43, 47, 53–56, 57–59, 61, B-2, B-4, C-3, D-2, E-2–E-3, G-1, G-4
Donations 1, 54, H-2
Duplication of benefits B-1–B-4, H-2

E

Educational facilities (PNP) A-1–A-6
Eligibility 3, 6–7, 13–19, 21, 24, 27, 30–31, 36, 40–41, A-1–A-6, G-4
Emergency 1, 2, 5, 8, 10, 19, 49, G-2
Emergency facilities (PNP) A-1–A-6
Emergency Protective Measures (Category B) ... 6, 10–11, 14–15, 38, 49, G-2
Emergency work 8, 10–11, 12–15, 36, 49, G-2
Environmental compliance 9, 41, 47, 62, E-1–E3, H-3
Environmental justice E-1
Equipment 1, 15–16, 26, 38, 42–47, 54, 56, A-4, B-1, B-3, F-1, F-19, G-3, H-2
Equipment rates 43
Expedited Payments 11, 22, 62, G-3

F

Federal Highway Administration (FHWA) 2, 15
Federal share 2, 6–7, 17, 25, 27, 29–30, 32, G-3–G-5
FEMA forms 4, 53, 55, F-1–F-20
FEMA-State Agreement 6, G-3
FEMA Web site .... i, 9, 13, 17, 43, 48, 56, 61, E-3, F-1, H-4
Fire Management (Category H) 17, H-1
Flood insurance 26, 54, B-1–B4, E-2, G-7
Force account equipment 1, 54, 56, F-1, F-19, G-2,
Force account labor 1, 36, 41, 54, 56, F-1, F-15, G-2

G

Gated communities .... *(see Homeowners' associations)*
Grass 17

H

Hazard mitigation 3–4, 8, 13, 21–23, 25–26, 28, 31, 33, 35, 41, 47,
54, 62, B-3, C-1–C-3, E-3, F-1, F-3, G-3
Historic Preservation 4, 8, 21–22, 25, 27–28, 31, 33, 35, 41, 47, 62,
C-2, D-1–D-2, E-1, F-1, F-14, G-7, H-3
Homeowners' associations A-5

I

Immediate Needs Funding 10–11, 22, 62, G-4
Improved project 23–25, 41, C-2, D-1, E-3
Indian Tribal government 7, 9, 17, H-1, H-2
Insurance 4, 8, 10–11, 21–22, 26, 31, 33, 35, 41–43, 46–47, 50, 54,
62, B-1–B-4, C-2, E-2, H-3
Irrigation 15
Irrigation facilities (PNP) A1–A-6

K

Kickoff Meeting 3, 7, 21–22, 30, 47, 62, G-4, H-3

L

Labor 1, 36, 41–43, 45, 54, 56, A-4, F-1, F-15, G-3, H-2
Large projects 29, 32–33, 42, 46–48, 53, 57–58, 63, G-4
Legal responsibility 6, 14, A-5
Location 22–23, 28, 35–36, 39, 53, 62, C-2, D-1, E-1–3, G-6, G-8

M

Major disaster 1–2, 5, 10, 17–19, 37, 49, 62, G-4
Management costs 17–18, H-2
Materials 1, 10, 42–45, 54, 56, F-1, F-16, H-2
Medical facilities (PNP) A-1–A-6
Mission Assignments 2, 6, H-1
Mitigation .... *(see Hazard mitigation)*
Mutual Aid Agreements 1, 55, G-5, H-1

N

National Flood Insurance Program (NFIP) 26, B-1–B-4
Natural Resources Conservation Service (NRCS) 2, 15
Non-critical PNP services A-4

O

Obligated funds 3, 33, G-5

P

PAC Crew Leader .... vii, 4, 7, 13, 21, 24, 28, 30–32, 36–37, 50, 61, B-2, B-4, C-1–C2, D-2, E-2, G-1, G-5
PA Group Supervisor 4
Parks, Recreational Areas, and Other Facilities (Category G) 16, 38, 49
Permanent work 3, 10–11, 15–17, 36, 49, 58, 62, A-4–A-5, G-5
Pre-application 3, 7, 9–11, 18, 22, 62, F-1, F-2, G-5, G-6
Pre-disaster design 15, 23, 25, 27, 40–41, C-1–2, E-1, E-3, G-5
Preliminary Damage Assessment (PDA) 2, 5–6, 10–11, 61, G-5
Private nonprofit (PNP) 3, 7–8, 17, A-1–A-6, F-1, F-13, G-5, G-6, H-1
Procurement .... *(see Contracts)*
Progress reports 57
Project Specialist 4, 16, 21, 30–32, 35, 37–38, 62, G-6
Project Worksheet (PW) .... *(see Subgrant Application)*
Public Entities 7

R

Reasonable cost 1, 31, 41–46, 61
Record keeping 8, 53–55, 57–59, 61
Repair 5–6, 15–17, 23–28, 40–41, 47, A-3, B-1, C-1–C-3, D-1, E-1, H-2
Repair vs. Replacement (50 Percent Rule) 23–24
Replacement 15–17, 23–25, 40–41, A-3, C-2, G-5, H-2
Request for Public Assistance (RPA) .... *(see Pre-application)*
Roads and Bridges (Category C) 15, 38, 49

S

Schools .... *(see Educational facilities)*
Scope of Work 22, 25, 29–33, 35, 39–41, 46–48, 49–51, 57–59, 62, B-4, C-2, E-3, G-8
Small Business Administration (SBA) A-4, A-6
Small project 29–32, 36, 42, 46, 48, 50, 53, 57–58, 62–63, D-2, G-7
Small project threshold 29, 46
Small project validation 30–32, F-1, F-11, H-3

Special considerations 3, 8, 21, 30–33, 35–36, 41, 47–48, 62, E-2, F-1, F-12, G-7
Special Flood Hazard Area (SFHA) B-1–B-2
State PA Representative 3–4, 5, 9, 32, G-1, G-7
Structures 13–16, A-1, D-1
Subgrant Application......................3, 7, 10–11, 18–19, 28–33, 35–41, 46–48, 49–51, 53–55, 62, A-4, B-1–B-3, B-4, C-1–C-2, D-2, E-2–E-3, F-1, F-4, G-6, G-8, H-3
Supplies 15, 43, A-1, H-2

T

Technical Specialist.................... 4, 21, 35, 41, 62, B-1–B-4, C-2, D-2, E-2, G-8
Time and materials contracts.. 45
Time extensions .... *(see deadlines)*
Time limits.. *(see deadlines)*
Trees 13, 17, H-2, H-4
Tribal government .... *(see Indian Tribal government)*

U

U.S. Army Corps of Engineers (USACE) .... viii, 2, 15
Utilities (Category F) 16, 38, 49, H-3
Utility facilities (PNP) A-1–A-6

V

Validation .... *(see Small project validation)*
Volunteers .... *(see Donations)*

W

Water Control Facilities (Category D) 15, 38, 49, A-2–A-3

www.ingramcontent.com/pod-product-compliance
Lightning Source LLC
Chambersburg PA
CBHW072034190526
45165CB00017B/870